GREAT POWER SECURITY COOPERATION

GREAT POWER SECURITY COOPERATION

Arms Control and the Challenge of Technological Change

David W. Kearn Jr.

LEXINGTON BOOKS
Lanham • Boulder • New York • London

Published by Lexington Books
An imprint of The Rowman & Littlefield Publishing Group, Inc.
4501 Forbes Boulevard, Suite 200, Lanham, Maryland 20706
www.rowman.com

Unit A, Whitacre Mews, 26-34 Stannary Street, London SE11 4AB, United Kingdom

British Library Cataloguing in Publication Information Available

Library of Congress Cataloging-in-Publication Data

Kearn, David W.
Great power security cooperation : arms control and the challenge of technological change / David W. Kearn Jr.
p. cm.
Includes bibliographical references and index.
ISBN 978-0-7391-8943-6 (cloth : alkaline paper) -- ISBN 978-0-7391-8944-3 (electronic) 1. Arms control--International cooperation. 2. Nuclear arms control--International cooperation. 3. Disarmament--International cooperation. 4. Great powers. 5. Security, International. 6. Technological innovations--Political aspects. 7. Military surveillance. 8. Arms control--International cooperation--History--20th century. 9. Nuclear arms control--International cooperation--History--20th century. I. Title.
JZ5625.K43 2015
327.1'74--dc23
2014042540

™ The paper used in this publication meets the minimum requirements of American National Standard for Information Sciences Permanence of Paper for Printed Library Materials, ANSI/NISO Z39.48-1992.

Printed in the United States of America

Contents

Acknowledgments

It has probably taken longer than it should have to transform my doctoral dissertation into a book. Nonetheless, I would first like to thank the members of my dissertation committee at the University of Virginia—Jeff Legro, John Owen, and David Coleman—all of whom provided numerous constructive comments and recommendations throughout the process and helpful ideas about moving forward with a book project. A special mention must be made for my dissertation adviser Dale Copeland who was enthusiastic, energetic, and supportive of the project even when I had doubts. He has been a great mentor and friend and I cannot thank him enough.

I have been extremely fortunate to have found giving and supportive people throughout my career and they have all directly or indirectly contributed to this project coming to fruition. Sincere thanks to Gordy Levin, Steve Walt, Lynn Davis, Herman Schwartz, David Waldner, Allen Lynch, Giulio Gallarotti, John Maurer, Robert Pfaltzgraff, and Andrew Winner. Other friends and colleagues who have helped along the way include Evan Montgomery, Stacie Pettyjohn, Abhishek Chatterjee, Zak Taylor, Robert Reardon, Markus Schiller, and Nick Miller. A special thanks to Keith Preble who compiled the index. I appreciate all of the help, support, and guidance from my colleagues in the Department of Government & Politics at St. John's University.

Thanks to Justin Race, Kathryn Tafelski, Brian Hill, and Elaine McGarraugh at Lexington Books/Rowman & Littlefield for all of their help in actually getting this book into print. They were a pleasure to work with.

Finally, I'd like to thank my parents, Maura and Dave, Sr. and my sisters, Kerry and Jessica, who have always been there for me. Without them, none of this would have been possible.

Introduction

This book seeks to explain the conditions under which states would most likely engage in a critical type of security cooperation: qualitative arms control.[1] Given an anarchic international system, when will states attempt to improve their security cooperatively rather than through unilateral means? Moreover, how can states improve their security—perhaps even decreasing the likelihood of war—by cooperatively prohibiting or limiting certain classes of weapons that would most likely increase their probability of victory in the event of a conflict?

To answer these questions, we must better understand how states view the potential military impact of technological change. A central premise of this project is that state perceptions relative the nature and characteristics of armaments are decisive in the formulation of their security strategy. In an anarchic system, states attempt to determine how existing assets contribute to their security. Perhaps more importantly, their leaders will seek to understand how technological change may be exploited to improve their capabilities through the development and deployment of new weapon systems and also how technological change can potentially decrease their security and create vulnerabilities that must be addressed should innovative adversaries develop and deploy similar weapons.

Thus is it important to understand how states approach decisions to develop, acquire, and deploy new weapons systems. How does technological change, specifically the expected military applications of that change, influence the security of states? Depending on their relative position in a given system, and taking into account the expected value of the new weapon system, their choices may vary. We would expect to see state military and political leaders assess tradeoffs between the perceived utility of new weapons systems and the potential impact of deploying those weapons may have on their security.[2] It is not difficult to envision scenarios in which the deployment of a new weapon system by one state would create such concern among its neighbors that the logical response by those neighbors would be to subsequently acquire and deploy similar assets, which could actually undermine the first state's security.[3] Assuming that states would see arms races as at least potentially dangerous, if not directly increasing the probability of war, cooperative attempts to limit or forego certain types of weapons may be seen as a desirable goal.[4] How do states address the challenges and opportunities presented by technological change within the larger context of developing their security strategies?

MILITARY EXPECTATIONS THEORY

The answers to these questions can be found in what I call Military Expectations Theory (MET). This theory begins with the idea that states will attempt to determine the impact of new military applications or weapon systems according to their potential to address existing strategic needs. The military and political leaders of a given state will seek to understand how new weapons systems fit into their existing military strategies and doctrines and, over time, how they might offer new means of improving the state's security and possibly lead to refinements or changes in strategy.[5] Innovative states will constantly work to improve their capabilities and also to tailor their doctrines and strategies to exploit those capabilities to their maximum potential. Where weapons systems are deemed to have the potential to be *decisive*, that is, where they may fundamentally alter the existing battlefield or provide dramatic new capabilities to the state deploying them, states will face strong pressures to develop and acquire them, though they will also fear their deployment by other states. In practical terms, the capabilities and postures of likely opponents will shape the perceived core missions and the requisite capabilities for any given state.[6] In this regard, the development of military expectations is an interactive process. Likely opponents and intervening factors like the geography of the likely battlefield will determine core missions. Discrete assets or systems of assets will be evaluated according to how effectively they allow states to carry out core missions, and thus assets that are deemed to be particularly effective will be highly prized. Conversely, assets that only marginally improve upon existing capabilities will be viewed with less enthusiasm by military leaders and, in turn, will be deemed less valuable. In the realm of arms control, we would expect that states would be more willing to bargain with these types of assets in order to improve mutual security precisely because the risks and costs associated with cheating may be more acceptable. Thus, by focusing on the potential utility and perceived benefits of various weapon systems themselves, MET provides a lens to

understand not only how states may seek to improve their military capabilities, but also how these capabilities may influence states to adopt more cooperative or competitive policies. MET explicitly links the perceived benefits of cooperation to the overall security strategy of the state.

Over time, the material elements of military power have changed, with varying degrees of influence on the strategies of states and their military and political leaders.[7] For much of history, military technology remained relatively stable over long periods of time. Breakthroughs in the waging of war (or so-called Revolutions in Military Affairs) have often been the result of the novel utilization of existing technologies or force structures within innovative doctrines.[8] However, in more recent cases, military revolutions have stemmed from major technological breakthroughs or the concurrent exploitation of several discrete technological breakthroughs to fundamentally change the nature of the existing battlefield. With the consolidation of political power inside the borders of the nation state and the subsequent and increasing bureaucratization of military affairs within it, states have become active in the search for and development of new technologies to improve their security. Innovations became more frequent, and as grand strategy subsumed military strategy as the central purview of statesmen, the stewardship of technological and industrial capacity capable of developing and producing new military applications became a critical focus.[9]

It seems clear that technological change has influenced state strategy and behavior over time. Military innovation is prized, potentially allowing states the capabilities necessary to execute more effective strategies.[10] Failure to exploit the benefits of new technology or the inability to adapt new applications to innovative doctrines can have drastic consequences in the face of an innovative enemy. Yet advantages can be fleeting, and once a discovery is made, technology, by its very nature, can allow for other states to "jump ahead" or at least "catch up" without having to follow the long and often painstaking road to innovation that a leader may have traveled. The increasing prevalence of a global arms market has further eroded longer-term advantages since states can purchase new military applications or the materiel, equipment and know-how necessary to develop those applications indigenously. Understanding how military technology, and changes in that technology, influence states' security strategies and behavior is thus an important component to understanding larger issues of international conflict and cooperation in world politics.[11] Technology has remained a relatively under-studied concept in international relations.[12] By focusing on the military impact of technological change and the expectations that states develop relative to that impact to contribute to their security, I submit that we can begin to answer important questions that have not been adequately addressed by existing theoretical approaches. Most importantly, we can explain the conditions under which states may seek and or agree to arms control initiatives to improve their security through cooperative rather than unilateral means.

THE IMPORTANCE OF THE QUESTION

While some would argue that the term "arms controller" became a pejorative term in policy circles during the late 1970s and 1980s, arms control represents an important concept in the both international relations theory and security studies. The central assumption of this study is that arms control is not an inherent "good" or an end unto itself. Instead, arms control is a focus of both academic and practical interest because it exists at the intersection of other important and interesting topics of study: international security, military strategy, technology and innovation, and international cooperation. More importantly, for arms control to be effective, it should enhance the security of the participants, contribute to and support their strategies, and realistically address issues in the context of technological change.[13] If it ignores any of these components, an arms control agreement will likely contain the seeds of its own failure. Yet if states are able to make credible bargains that restrain their own capabilities to harm each other and thus create and sustain a relatively benign security environment, we would expect that arms control or arms limitations and/or reductions would be an important method to cooperatively address international insecurity challenges. However, both historical record and the theoretical canon reflect the practical difficulty of achieving sustainable arms control agreements. In an uncertain, anarchical environment, we would expect that cooperation in the security realm would be very difficult, if not impossible. Nonetheless, we have seen examples of successful arms control and those successes have significantly influenced the subsequent diplomatic relations of the relevant states. Discerning the causes of success and failure in the realm of arms control has both important theoretical and practical consequences that clearly make such a study worthwhile and valuable.

The role of armaments and military technology more generally, in international relations has been an issue of theoretical and practical interest since at least the turn of the last century. As technology provided states with more powerful weapons and war became increasingly total in nature, statesmen and scholars sought to focus on armaments, and restraints on armaments, as possible means to dampen or minimize the potential for conflict. International disarmament conferences attempted to grapple with the issues of restraining certain weapons viewed as particularly dangerous or destabilizing prior to both of the World Wars.[14] The results were disappointing, and skeptics have long argued that statesmen and leaders, or the deeper political and societal forces that drive them, are responsible for war. Armaments are merely tools that they use. Yet the underlying notion that a deep connection existed between the militarization of states and the occurrence of war persisted.[15] Soon after the detonation of the first atomic weapons, the U.S.government sought to develop means for cooperative international control of nuclear materials. While these and other early initiatives failed, the history of the Cold War was marked by a variety of arms control negotiations, some successful, some not.[16] The objectives seemed to turn to managing the arms competitions between the two superpowers rather than eliminating or prohibiting certain classes of weapons, which defined many of the earlier conventional attempts. Nonetheless, the fundamental idea that agreement on managing or limiting certain weapons systems could contribute to a more stable and peaceful environment persisted, even as deeper political differences remained. In the most optimistic of views, perhaps successful arms control could create favorable security environments that would contribute to the resolution of political differences. Conversely, unfettered arms races could lead to increased tensions, setting the stage for crisis and potentially war. Thus, scholars have attempted to investigate the relationship between arms races and war, and while both phenomena may be driven by underlying political conflicts, the concept of cooperative international arms control as a potential remedy has enjoyed a troubled theoretical and practical history.[17]

Far from being mere questions of academic interest, the issues explored in this project are critical to understanding current security challenges. The findings of this study should have policy relevant implications that will be useful for understanding the strategies of states facing technological breakthroughs that could dramatically change their own security. The growing security competition between the United States and China will likely reflect these types of dynamics. China has already moved to develop weapons systems that exploit cutting edge technologies to exploit perceived weaknesses in the U.S. defense posture, such as offensive cyber-warfare capabilities, anti-satellite (ASAT) weaponry, and a high-tech missile force that threatens to turn U.S. naval assets into liabilities.[18] We would expect that powers like China and Russia would continue to view U.S. antiballistic missile (ABM) systems with great concern and, more importantly, to undertake measures to address them.[19] Potentially revolutionary technologies with military applications like cyber-warfare, hypersonics, and nanotechnology may also spark new rounds of competition. Understanding how states perceive the contributions of new military applications to their overall security, will go a long way in predicting their decisions to acquire and deploy new weapons as well as their willingness to cooperatively limit or restrain them through arms control.

METHODOLOGY

This study employs the methods of structured, focused comparison of cases, in-case variation, and process tracing.[20] The historical case studies that make up the empirical chapters have been developed to test the strength of the MET and its subordinate hypotheses against theoretically and historically informed competing alternatives, and have been evaluated against a set of critical questions.[21] This structure of the investigation will "guide and standardize data collection, thereby making systematic comparison and accumulation of the findings of the case possible."[22] This explicit clarity with regard to measurement and the coding of variables will improve the study's contribution to the theoretical discussions and debates considered within. Moreover, I will focus only on certain specific aspects of the cases to draw out the importance of the expectations of leaders with regard to technological change. Many other interesting aspects of the cases, from a historical or theoretical view, will therefore be set aside.

DEVELOPING CASES

After developing a research design and creating the subsequent critical questions, hypotheses and alternative hypotheses based upon existing International Relations (IR) theories to test, the central task was to find cases that are comparable.[23] For example, "cooperative arms control initiatives" include a wide variety of political endeavors. There have been various attempts to develop accepted rules of warfare, while other agreements focused on the geographic prohibitions of certain weapons systems. Other demilitarization arrangements have been created to support political settlements at the conclusion of violent conflicts. The subset of cases examined in this study possesses a common underlying theme. The cases that have been chosen for this study reflect a specific form of arms control: qualitative arms control. In each of these cases, states attempted, and in some cases quite presciently, to grapple with the potential security problems presented by certain kinds of weapons systems.[24] The focus on types of weapons systems is the common thread between them. The ultimate choice of cases reflects a combination of John Stuart Mill's Methods of Difference and Similarity.[25] Focusing on the two independent variables, expectations of the military impact of new technologies and prevailing means of surveillance, I attempted to find cases that were in most other measures, similar, but where the independent variable varied significantly. The first two cases involved several great powers struggling to develop limitations on naval weapons and land weapons, respectively, while the third and fourth cases involved the Cold War superpowers negotiating limits to nuclear weapons. In both cases where the independent or study variable, Expected Military Impact of Technological Change was "low," leaders were able to successfully negotiate cooperative arms limitation agreements, where when this value was "high," leaders were unsuccessful in achieving a cooperative, negotiated settlement. Interestingly, in the two "successful" cases, Washington and SALT I, there was also significant in-case variation across different weapons systems, and while agreement allowed for the limitation of primary assets, other weapons could not be cooperatively restrained. In each case, leaders were presented with the opportunity to limit or prohibit certain classes of weapons to avoid potential arms races and thus decrease the probability of war. Whether they sought limitations or could ultimately forge an agreement is the focus of the empirical investigation.

IN-CASE VARIATION AND PROCESS TRACING

An additional benefit to utilizing cases that focus on specific attempts to address the challenges created by specific technologies and their military and security implications over a period of time is the ability to increase our observations as well as investigate significant in-case (or "within-case") variation that should provide greater explanatory leverage and allow us to further isolate and explicate the relevant causal mechanisms in against a relatively broad and diverse set of contextual factors.[26] In-case variation that allows for differing measures on the dependent variable provides us with opportunities to make stronger inferential claims and can highlight the influence of antecedent conditions or intervening variables that may have a greater or lesser influence on outcomes and allowing for control.[27] Additionally, as the cases are constructed, we may also uncover further avenues to explore that impact of the independent variables and causal mechanisms that constitute the theory presented in this study.

The process-tracing approach "involves both an attempt to reconstruct actors' definitions of the situation and an attempt to develop a theory of action."[28] By using documentary and archival evidence and other primary empirical sources it should be possible to construct a picture of the thinking of key decision-makers prior to the actual negotiations. This key task allows us to avoid the problem of tautology, simply inputing the motivations and intentions of the leaders simply by viewing the subsequent political outcomes. It is then possible to draw a chain of events reflecting the political processes involved in that led to the eventual outcome.[29] We should thus be able to understand why they came to agree upon a given strategy before it is actually implemented.[30] Moreover, in employing process tracing, we seek to illustrate that the military and political leaders of states do indeed act as if they are acting according to the behavior predicted by the hypotheses presented.[31] Specifically, by utilizing the documentary evidence of leaders grappling with the problems of developing strategies in the face of innovation and adapting new technologies to their strategic needs, we should be able to piece together a clear narrative of how they approach possible arms control initiatives.

Process-tracing allows us to focus squarely on the independent variables. These expectations can be described in rich detail based on archival materials, memoirs, and official documentary evidence. In many cases there may be debates among military and political leaders concerning the expected impact of new military applications, particularly in new innovative weapons that may require investments of limited resources from existing deployed or planned weapons based on mature technologies. Similarly, process tracing will be employed to accurately describe the perceptions of military and political leaders regarding the effectiveness of prevailing means of surveillance, the second independent variable. The critical question we seek to answer is how confident are military and political leaders that prevailing surveillance capabilities can detect cheating by an adversary before it was too late to take effective counter measures? As with expected impact of new military technologies, this question may be the subject of intense debate within policy circles, but this should allow us to distill a relatively clear picture of the views of relevant decision-makers concerning their state's ability to detect cheating subsequent to an arms control agreement. It is important to note that one critical factor that may arise is the perceived need for greater surveillance and monitoring capabilities than currently exist. The construction of an expansive, intrusive regime may be deemed necessary to effectively safeguard an agreement, but it is unlikely to prove capable of providing the confidence associated with unilateral national means of surveillance, contributing to a breakdown of cooperation. Ultimately, process-tracing allows us to investigate these perceptions and concerns and accurately describe their influence on the eventual policy decisions.

We would expect to see that the introduction (the decisions to invest in research and development, acquisition, and deployment) of the military applications of new or emerging technologies entails a political process.[32] Leaders, in many cases, must be brought to see the potential advantages of a new application, often with some sort of tangible results that can be then used to support requests for the necessary investment. As Barry Posen has explained, it is often those technological applications that have exhibited some level of success in a real-world test setting that are most likely to be effectively integrated in a timely manner.[33] The use of surrogates or developmental prototypes on an actual battlefield or in controlled war game settings may provide the crucial evidence needed to impress higher level military leaders of the value of a particular weapon system. Similarly, attaché reports or other sources of external information that reflect the success of a new application in a conflict can heavily influence the attitudes of interested observers.

THE EVIDENCE

While the documentary evidence will be taken from sources that are created within a given state, if we find that they reflect the rational decision-making process that would be predicted by structural theories, we can reasonably conclude that the leaders are acting as if they are responding to structural constraints and opportunities. In terms of measurement, we would expect to see, reflected in the documentary evidence as well as secondary sources and subsequent analyses, discussion and argument supporting the adoption and deployment of new military applications and their potential value to a state's given security needs. Where a new application or system of applications is perceived as being potentially decisive, the documents should clearly reflect that military and political leaders actively sought to acquire and deploy these systems. Conversely, where less evidence exists of military leaders actively pursuing new the acquisition of new systems, we would not expect that they foresee their utility in a future conflict to be high, or their potential impact decisive. More usefully, evidence that leaders viewed existing assets as mature and unlikely to change—and the impact of technological change to be minimal over the near-to-medium term—subsequent documentation should reflect less enthusiasm concerning their acquisition or deployment. A third possibility is that state leaders may be unsure of the potential impact of a given application of a new technology in a relevant time frame. With this variation of the independent variable, we would expect to see variation on the dependent variable—the state's decision to engage in cooperative arms control initiatives. The ultimate success of a given arms control or disarmament negotiation will necessarily involve the cooperation of the various actors involved and negotiations could fail even in the presence of genuine willingness by one or more of the partners. This project is thus limited to explaining how technological expectations influence the strategy of states approaching cooperative initiatives. When will states seem to exhibit deadlock preferences, preferring no settlement even if this means that a potential arms race ultimately undermines their security? Conversely, when will states view cooperative arms control as a preferred strategy over unilateral arming? These questions remain within the realm of a state foreign policy or grand strategy, but if they support structural realist predictions then we can, with some certainty, affirm that states act according to structural, rather than domestic-level imperatives.[34] In turn, the results of the study should help us better understand the conditions under which we may expect international arms control cooperation to succeed.

In examining cases that span significant periods of time, rather than just discrete diplomatic events, the study is able to address two important factors: the influence of timing, and specifically the differing impact of new or emerging technologies versus known, tested applications of mature technologies on expectations of military and political leaders, and the respective positions of states in terms of relative power or capabilities. This allows us to truly isolate the impact of expectations and technological change against other concerns like relative power differentiations or perceptions of leaders and laggards and also provides us with a means for explaining how these factors may interact in terms of cooperation or competition. To a lesser degree, looking a relevant time periods may allow us to better understand whether relevant states were truly security seekers or perhaps revisionist, in which case arms control would likely be a futile endeavor.

Moreover, in constructing cases that focus on attempts to address the challenges created by specific technologies and their military and security implications over a period of time, we are able to increase our observations as well as investigate significant in-case variation that should provide further explanatory leverage and allow us to further isolate and explicate the relevant causal mechanisms in against a relatively broad and diverse set of contextual factors.

THE PLAN OF THE STUDY

Chapter 1 lays out the core of the Military Expectations Theory, including the assumptions, model, variables, and hypotheses that will be examined in this project. Military technology has dramatically shaped the nature of the battlefield over time, and while wars may ultimately be caused by political conflicts between states, the utilization of new technological applications can often have significant effects on their outcomes. Knowing this, states existing in anarchy have developed bureaucratic and technical capacities to shape and exploit technological change, with the ultimate goal of developing and obtaining new capabilities to improve their security. Technology is not a linear property, however. Technology can be manipulated to a certain degree, but there are limits to what states can achieve—despite their best efforts—in the short term. States understand both their own relative positions and that of their potential adversaries in terms of technological efficacy as well as their calculations of the distribution of relative power in a given system. The expectations that leaders must develop with regard to the military impact of new technological applications will shape the nature of their strategies. We can then understand how the perceived value of a new system will influence the state's willingness to agree arms control. Similarly, while expectations may drive states to seek out or respond favorably to arms control initiatives, the second independent variable, perceptions of prevailing means of surveillance which will shape the ultimate outcome of attempted cooperation. The chapter also constructs a number of alternative hypotheses based on other theoretical traditions to set up the competitive theory testing that I will undertake in the historical case studies.

The second chapter examines the history of the Washington Naval Conference. Focusing on the limitation of the strategic weapon of the era—the dreadnought battleship—the United States, Great Britain, and Japan succeeded in dampening a potential conflict in the western Pacific and improving the security of the principals. The static nature of battleship technology and the potential for other emerging technologies to limit the effectiveness of capital ships as primary weapons systems allowed the United States and eventually its partners to agree to significant limitations and a successful treaty agreement, though other weapons were not effectively controlled. Chapter 3 examines the failure of the World Disarmament Conference at Geneva (1931–1933). In examining this case, however, it is important to consider the work of the Preparatory Commission of the World Disarmament Conference which was created under the League of Nations to address the groundwork of the conference prior to the meetings of high-level diplomatic representatives. Starting in 1926, the Preparatory Conference sought to delineate the differences between offensive and defensive weapons and structures to control the former. Thus, despite nearly a decade of effort, leaders were unable to cooperatively limit or abolish aggressive armaments and avoid overturning a defensive security environment.

Chapter 4 focuses on the attempts of the United States to avoid an atomic arms race after World War II. Specifically, the Baruch Plan, which sought to create international control over atomic energy under the auspices of the United Nations, will be examined. While the United States and the Soviet Union failed to reach an agreement on a working structure for international control of atomic energy, the American proposal was revolutionary even by present standards. However, the absolute nature the atomic bomb insured that it was simply too great a challenge for any meaningful cooperative outcome. The strategic competition was quite different in the late 1960s and early 1970s when the United States and Soviets were able to come to agreement in the SALT I accords, which are examined in chapter 5. Parity had more or less been achieved by the Soviets. More importantly, acceptance that the central assets of both superpowers' strategic arsenals, Intercontinental Ballistic Missiles (ICBMs) represented a technology that was going to be fairly static for the foreseeable future allowed for possible agreement. Moreover, an emerging technology that was seen to likely provide dubious results at best, antiballistic missile systems (ABM), were banned in an effort to maintain the "delicate balance of terror" and reinforce the assured destruction concept that had come to define deterrence by the 1970s. Nonetheless, the challenge of multiple independent reentry vehicles (MIRV) remained unresolved. A brief conclusion will consider the predictions and implications for international relations theory and security studies as well as several potential areas of security cooperation in the current international security environment.

NOTES

1. Under certain conditions, we would expect that arms control would be a fruitful means to improve cooperation. See, for example: George W. Downs, David M. Rocke, and Randolph M. Siverson, "Arms Races and Cooperation," *World Politics* 38, no. 1 (1985); Charles L. Glaser, "Realists as Optimists: Cooperation as Self-Help," *International Security* 19, no. 3 (1994/95); Robert Jervis, "Arms Control, Stability, and Causes of War," *Political Science Quarterly* 108, no. 2 (1993); Thomas C. Schelling and Morton H. Halperin, *Strategy and Arms Control* (Washington, DC: Pergamon-Brassey, 1985).

2. This is the crux of the security dilemma. Rational, security-seeking states should be aware of the consequences of their choices. See Robert Jervis, "Cooperation under the Security Dilemma," *World Politics* 30, no. 2 (1978).

3. See Charles L. Glaser, "The Security Dilemma Revisited," *World Politics* 50, no. 1 (1997), especially 173–83.

4. Glaser, "Realists as Optimists," 64–67.

5. On the interaction of technology and doctrine, Stephen Van Evera, *Causes of War: Power and the Roots of Conflict* (Ithaca, NY: Cornell University Press, 1999), especially 161–62. Barry Posen explains the importance of "political-military integration" and place innovation as a primary component in creating and maintaining that integration over time. See Barry Posen, *The Sources of Military Doctrine: France Britain and Germany between the World Wars* (Ithaca, NY: Cornell University Press, 1984), especially 25–40.

6. The approach utilized in this study attempts to take a more fine-grained view of "power" by focusing on military capabilities. Van Evera develops this concept in his *Causes of War: Power and the Roots of Conflict*, 7–11. This focus on military assets also follows from the approach Charles Glaser developed in his "contingent" realism in "Realists as Optimists," 58. Also see Charles L. Glaser, "Political Consequences of Military Strategy: Expanding and Refining the Spiral and Deterrence Models," *World Politics* 44, no. 4 (1992).

7. On the changing nature of warfare, see Bernard Brodie and Fawn Brodie, *From Crossbow to H-Bomb* (New York: Dell, 1962); William H. McNeill, *The Pursuit of Power: Technology, Armed Force, and Society since A.D. 1000* (Chicago: University of Chicago Press, 1984); Martin Van Creveld, *Technology and War: From 2000 B.C. to the Present* (New York: Free Press, 1989).

8. For definitions of "Military Revolutions," see MacGregor Knox and Williamson Murray, eds., *The Dynamics of Military Revolution 1300–2050* (Cambridge: Cambridge University Press, 2001); Andrew F. Krepinevich, "Cavalry to Computer: The Pattern of Military Revolutions," *National Interest*, no. 37 (1994).

9. On the components and requisites of grand strategy, see, for example, Paul M. Kennedy, *Grand Strategies in War and Peace* (New Haven, CT: Yale University Press, 1991); Posen, *The Sources of Military Doctrine*, 13.

10. John J. Mearsheimer, *The Tragedy of Great Power Politics* (New York: W. W. Norton, 2001), 166–67.

11. Barry Buzan and Eric Herring, *The Arms Dynamic in World Politics* (Boulder, CO: Lynne Rienner, 1998).

12. Notable exceptions include Geoffrey Herrera, *Technology and International Transformation: The Railroad, the Atom Bomb, and the Politics of Technological Change* (Albany: State University of New York Press, 2006).

13. Thomas Schelling and Morton Halperin, *Strategy and Arms Control* (Washington, DC: Pergamon-Brassey, 1985).

14. On conventional arms control in the early twentieth century, see Coit D. Blacker and Gloria Duffy, eds., *International Arms Control: Issues and Agreements*, 2nd ed. (Stanford, CA:

Stanford University Press, 1984); B. J. C. McKercher, ed., *Arms Limitation and Disarmament: Restraints on War, 1899–1939* (Westport, CT: Praeger, 1992).

15. Joseph Maiolo, *Cry Havoc: How the Arms Race Drove the World to War, 1931–1941* (New York: Basic Books, 2010).

16. Excellent evaluations of Cold War arms control negotiations can be found in George Bunn, *Arms Control by Committee: Managing Negotiations with the Russians* (Stanford, CA: Stanford University Press, 1992); Albert Carnesale and Richard N. Haass, eds., *Superpower Arms Control: Setting the Record Straight* (Cambridge: Ballinger, 1987); Alexander L. George, Philip J. Farley, and Alexander Dallin, eds., *U.S.-Security Cooperation: Achievements, Failures, Lessons* (New York: Oxford University Press, 1988).

17. On the nature of arms races, see, for example, Charles L. Glaser, "The Causes and Consequences of Arms Racing," *Annual Review of Political Science* no. 3 (2003); Colin S. Gray, "The Arms Race Phenomenon," *World Politics* 24, no. 1 (1971); Colin S. Gray, "The Urge to Compete: Rationales for Arms Racing," *World Politics* 26, no. 2 (1974); Samuel P. Huntington, "Arms Races: Prerequisites and Results," in *The Use of Force*, ed. Robert J. Art and Robert Jervis (Lanham, MD: University Press of America, 1993). On the relationship between arms races and war, see Paul F. Diehl, "Armaments without War: An Analysis of Some Underlying Effects," *Journal of Peace Research* 22, no. 3 (1985); Paul F. Diehl, "Arms Races and Escalation: A Closer Look," *Journal of Peace Research* 20, no. 3 (1983); Paul F. Diehl, "Arms Races to War: Testing Some Empirical Linkages," *Sociological Quarterly* 26, no. 3 (1985); Paul F. Diehl, "Ghosts of Arms Control Past," *Political Science Quarterly* 105, no. 4 (1990); Paul F. Diehl and Mark J. C. Crescenzi, "Reconfiguring the Arms Race-War Debate," *Journal of Peace Research* 35, no. 1 (1998); Susan G. Sample, "Arms Races and Dispute Escalation: Resolving the Debate," *Journal of Peace Research* 34, no. 1 (1997); Susan G. Sample, "Furthering the Investigation into the Effects of Arms Buildups," *Journal of Peace Research* 35, no. 1 (1998); Michael D. Wallace, "Arms Races and Escalation: Some New Evidence," *Journal of Conflict Resolution* 23, no. 1 (1979); Michael D. Wallace, "Comments on the Articles by Sample and Diehl and Crescenzi," *Journal of Peace Research* 35, no. 1 (1998).

18. On Chinese military modernization, see Thomas J. Christensen, "Posing Problems without Catching Up: China's Rise and Challenges for U.S. Security Policy," *International Security* 25, no. 4 (2001); William Martel and Toshi Yoshihara, "Averting a Sino-U.S. Space Race," *Washington Quarterly* 26, no. 4 (2003); Brad Roberts, Robert A. Manning, and Ronald N. Montaperto, "China: The Forgotten Nulclear Power," *Foreign Affairs* 79, no. 4 (2000).

19. On ABM issues, see, for example, James M. Lindsay et al., "Limited National and Allied Missile Defense," *International Security* 26, no. 4 (2002).

20. On the development of case studies see Stephen Van Evera, *Guide to Methods for Students of Political Science* (Ithaca, NY: Cornell University Press, 1997), 49–89. Alexander L. George and Timothy J. McKeown, "Case Studies and Theories of Organizational Decision Making," in *Advances in Information Processing in Organizations*, ed. Robert F. Coulam and Richard A. Smith (Greenwich, CT: JAI, 1985).

21. Alexander L. George and Andrew Bennett, *Case Studies and Theory Development in the Social Sciences* (Cambridge: MIT Press, 2004), 69.

22. Ibid., 67.

23. Gerardo L. Munck, "Tools for Qualitative Research," in Henry E. Brady and David Collier, *Rethinking Social Inquiry: Diverse Tools, Shared Standards* (Lanham, MD: Rowman & Littlefield, 2004), 105–21.

24. Timothy J. McKeown, "Case Studies and the Limits of the Quantitative Worldview," in Brady and Collier, *Rethinking Social Inquiry*, 139–67.

25. On Mill's methods, see Van Evera, *Guide to Methods for Students of Political Science*, 23–25.

26. George and Bennett, Case *Studies and Theory Development in the Social Sciences*, 151–81; Van Evera, *Guide to Methods for Students of Political Science*, 61–63.

27. Munck, "Tools for Qualitative Research," 116–18.

28. George and McKeown, "Case Studies and Theories of Organizational Decision Making," 35. See also Van Evera, *Guide to Methods for Students of Political Science*, 64–66.

29. Andrew Bennett, "Process Tracing and Causal Inference," in Henry E. Brady and David Collier, *Rethinking Social Inquiry: Diverse Tools, Shared Standards*, 2nd ed. (Lanham, MD: Rowman & Littlefield, 2010), 207–19.
30. George and Bennett, *Case Studies and Theory Development in the Social Sciences*, 6.
31. Van Evera, *Guide to Methods for Students of Political Science*, 29.
32. Stephen Peter Rosen, *Winning the Next War: Innovation and the Modern Military* (Ithaca, NY: Cornell University Press, 1991), 2–3.
33. Posen, *The Sources of Military Doctrine*, 55–57.
34. Richard Rosecrance and Arthur A. Stein, *The Domestic Bases of Grand Strategy* (Ithaca, NY: Cornell University Press 1993).

Chapter 1

Understanding Security Cooperation and Technological Change

As a primary means of cooperation in the security realm, arms control has a troubled history. Even under relatively favorable political conditions in the past, the achievement of lasting arms control agreements has been rare. Nonetheless, there have been important successes which have allowed states to improve their security by cooperatively restraining their military capabilities and forgoing potential opportunities. How can we understand the conditions under which arms control is likely to succeed or fail? In an anarchic world, how can great powers achieve successful negotiated settlements that are viewed to improve their security and maintain those settlements over time?

This book seeks to provide answers to these questions. By focusing on the influence of technological change on state security strategies, it is possible to understand the conditions under which security-seeking great powers may seek out and/or engage in cooperative arms control to improve their security and avoid potentially self-defeating or suboptimal outcomes that would result from unilateral policies. Precisely because the costs of defection and the risks associated with arms control are grave, states must be confident that cooperation will not lead to the creation of vulnerabilities that could threaten their survival. This study argues that the influence and interaction two critical technologically influenced factors—the expected military impact of technological change (or the expected capabilities offered by the development and deployment of new weapons systems), and the prevailing forms of surveillance available to states—will determine whether states will view arms control initiatives favorably, and whether these initiatives ultimately succeed or fail. The first variable shapes the receptiveness of states to limitations or reductions in their stocks of specific weapons if doing so would improve their security, while the second determines whether an agreement can be sustained.

This chapter sets out the core logic of Military Expectations Theory (MET). The first section will consider the problem of security cooperation and qualitative arms control in the context of mainstream international relations theory. Neither defensive realist approaches, which do incorporate the influence of military technology, nor offensive realist analyses, which are less explicitly concerned with technology, have adequately addressed the impact of technological change on state security strategies. MET seeks to synthesize these approaches and provide a more comprehensive explanation of the influence of technological change on state behavior and the likelihood of successful cooperation, thus filling an important gap in the literature. The core of MET will be developed in the second section. The two independent variables will be explained: the military expectations of technological change, which will condition leaders' views on cooperation, and perceptions of prevailing means of surveillance, which will determine whether leaders can credibly commit to cooperative initiatives. In addition, two key parameters—the nature of the security environment, and the timing of military innovation which may create a leading state—will be examined. The third section briefly considers the expected benefits of security cooperation. The final section will develop a set of testable hypotheses derived from the logic of MET as well as some alternative hypotheses based upon other theoretical approaches. These will be tested against the empirical evidence in the book's historical chapters.

UNDERSTANDING SECURITY COOPERATION

Major theories of international relations provide incomplete and unsatisfactory explanations for cooperation in the security realm. For example, offensive realism would argue that it is impossible for states to cooperate because the costs of defection are so great.[1] If we witness cooperation in the empirical world, these scholars argue that it is only in a context in which states are unlikely to pay high costs for defection—or where it's "easy." However, this approach downplays or ignores the risk and potential costs of consistent hardline policies.[2] Defensive realism, based on the logic of the security dilemma and utilizing the offense-defense balance, presents a more optimistic picture of international relations and acknowledges the potential benefits of cooperation, but it is limited in its ability to explain variations in patterns of cooperation, particularly over extended periods of time.[3] Institutionalist scholars focus on the importance of regimes in facilitating collective action and supporting cooperation by providing surveillance and information on compliance to deter defection.[4] However, this literature has been notably limited in security affairs.[5] The challenge for this approach is to explain why states would cooperate in the first place, where costs of defection in the security realm are potentially so high.[6] Moreover, in an anarchic international system, it is unclear that a rational, security-seeking state could ever entrust its security to an international institution or regime. The theory of military expectations developed in this chapter directly addresses these existing limitations in major theoretical approaches and presents a more comprehensive explanation for cooperation in the security realm.

SECURITY-SEEKING STATES IN ANARCHY

Offensive realism focuses primarily on the distribution of relative power in a system to explain state strategy and behavior.[7] However, it is the notion that states will (and should) seek to maximize their relative power that sets offensive realism apart as a theoretical approach to understanding international relations.[8] The implications of this assumption is that states will rarely seek or even consider cooperation with other states because of fears of cheating and relative gains concerns.[9] This has little direct applicability to questions of arms control cooperation, aside from perhaps concerns about quantitative arms limitations. But the real insight that offensive realism offers is the fear of future intentions that drives much of the hardline policies that it predicts. States must be concerned about their future. While the presence of an imminent threat obviously forces the leaders of states to focus their energies on short-term concerns, most leaders will also be concerned with medium- and even long-term concerns. In peacetime, this may boil down to a tradeoff of investment between readiness and development as states grapple with the uncertainties of their respective security environments, in terms of both time and space. The model of state behavior considered in this study similarly takes the notion of state fears of the future seriously, and thus also views the notion of time and changes over time as critical to perceptions and decisions of state leaders.

Understanding the potential downside of overly aggressive behavior, defensive realists assume that security-seeking states are less concerned with the accumulation of relative power, even if it is a means to an end, and more squarely focused on acquiring the capabilities deemed necessary to provide for their security.[10] Rather than seeking to maximize power, states seek to maximize their security, and cooperation can prove a useful means to achieve that end goal. While states are likely to take active measures to deter states that are viewed to possess greedy motives, they also understand the outcomes of crises, arms races, and wars are highly uncertain and that they could indeed lose. Facing such a prospect, states have a motivation to attempt to discern fellow security seekers from greedy states, and cooperative initiatives may provide such a tool.

This study thus synthesizes offensive and defensive realist approaches, taking concerns for the future seriously, while also adopting most defensive realist assumptions about the motivations and preferences of states and their leaders.[11] However, the model developed in this study presents a more realistic and accurate presentation of states (and their military and political leaders) and their concerns about present and future security. The next section assesses the most developed theoretical concept for explaining cooperation in security realm: the offense-defense balance. It will also address the criticisms of offense-defense theory derived from other theoretical approaches in order to introduce the key elements of a refined, technologically driven approach which I term Military Expectations Theory, to understand the conditions under which cooperation is likely to succeed and, conversely, where cooperation is likely to fail and states are likely to adopt more competitive policies.

OFFENSE-DEFENSE THEORY AND TECHNOLOGICAL CHANGE

Much of the defensive realist literature incorporates the offense-defense balance as an important variable for understanding the likelihood of conflict or cooperation between dyads of states or within a given system of states. Moving away from more expansive definitions of the offense-defense balance and focusing on a core concept of technology and geography, it is possible to characterize the security environment facing states and their leaders in terms of *defensive* or *offensive* dominance or an effectively *neutral* environment.[12] While scholars have provided differing ways to measure the extent to which a system is offense or defense dominant, in the most straightforward explanation the balance determines the expected costs of military operations, all else being equal.[13] Where a security environment is defense dominant, effective defenses are inexpensive and the execution of offensive operations is prohibitively costly. Conversely, where geography and technology combine to create a highly offensive environment, defenses are exorbitantly expensive and offensive operations are inexpensive and highly effective.[14]

The offense-defense balance thus remains a structural (though often considered more "fine-grained") variable, creating constraints or opportunities for states at any given point in time.[15] Rational military and political leaders should seek to understand the nature of the existing balance and construct their forces accordingly. It is important to note that this discussion does not imply that the offense-defense balance determines military outcomes. It is a structural constraint, confronting states and their leaders with a set of restrictive or permissive conditions that will influence the ultimate success of their strategies. At any time, the effects of the balance could be mitigated through tactical innovations or overcome through the commitment of overwhelming power. Yet whether soldiers develop new techniques for penetrating static defenses under fire, or whether a great power overruns the well-constructed defenses of a smaller power at great cost, the offense-defense balance shapes but does not determine the outcomes.[16]

In peacetime, if states are able to distinguish offensive capabilities from defensive capabilities and determine which are more effective, then leaders may be able to signal peaceful intentions to other states using their military policies.[17] In doing so, they may be able to improve their security through cooperative, rather than competitive, gestures and dampen potential rivalries or even escape the security dilemma. In a defense dominant world, where technology allows states to effectively defend themselves without threatening others, states should be able to cooperate with little fear of the potential costs defection by an adversary. Similarly, in an environment that is effectively neutral in offense-defense terms, states may be able to craft arms control agreements, particularly if offensive and defensive weapons can be distinguished.[18] Conversely, if states have difficulty distinguishing what type of system they face, they may be resistant to engaging in cooperation that could leave them vulnerable in the event of defection.[19] However, arms control initiatives, would seemingly be particularly useful to a state's security strategy in a world where states are indeed capable of determining the nature of the system based upon prevailing military technologies.[20] In an offense dominant system, states must prepare for the worst and cooperation—while potentially valuable for rational, security-seeking states—is exceedingly difficult because of the high potential costs associated with cheating. Small advantages can have large implications, and states must be on guard against first-strikes.[21]

While other criticisms have been leveled against it, a major shortcoming of the offense-defense balance is it is overly (and perhaps artificially) static.[22] If the two major factors determining the nature of the balance are geography and technology, for any given system (or for a specific dyad) of states, only one of these can actually change. Geography is constant, but technology is an inherently dynamic property; to some extent it is *always* changing. Thus, when we boil the offense-defense balance down to its core, it is essentially a static snapshot of the cumulative influence of prevailing military technologies at a discrete point in time, accounting for prominent geographic influences.[23] However, when we move away from the discrete point in time, there is potential for the balance to change. For example, even if a security environment is squarely defense dominant at time t_1, technological change may soon provide applications to innovative states that effectively overturn that environment and shift the balance to the offense by time t_2. (See figure 1.1.)

It is important to recognize that technology does not simply change over time in a predictable way.[24] Where technological change is perceived to be rapid, providing new weapon systems and shifting the balance in significant but unpredictable ways, we would expect cooperation to be virtually impossible, and that rational, security-seeking states would have little reason to incorporate notions of the balance into their military plans precisely because it is inherently unstable. If technological change could be expected to constantly offer new and decisive weapons systems then the offense-defense balance would have almost no practical or analytical utility.[25]

TECHNOLOGICAL OPPORTUNISM

This is essentially the argument that Kier Lieber makes in his book *War and the Engineers*.[26] Introducing the concept of "technological opportunism" Lieber argues that technology serves state political interests and the purposeful development of new capabilities to better achieve a state's vital objectives is precisely the type of activity that one would expect in a competitive, self-help environment.[27] Lieber dismisses the importance of existing technological realities at the systemic level (i.e., the offense-defense balance), pointing out that military and political strategies chosen and executed by states do not

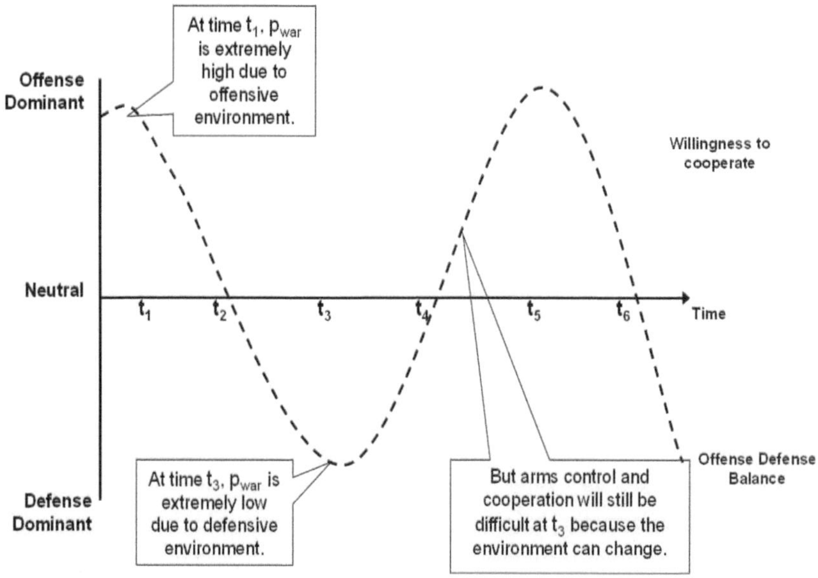

Figure 1.1 Dynamic Technology and the Limits of Offense-Defense Balance Influence on States

necessarily follow directly from the constraints and opportunities created by the offense-defense balance at any point in time.[28] Thus, Lieber's critique of offense-defense theory is an important one: because technology is an inherently dynamic property, the offense-defense balance is ephemeral at best representing the technological reality at a discreet point in time. However, Lieber goes too far in incorporating an overly linear and rapidly changing depiction of technology into his model.

If the nature of technological change was such that it constantly provided new applications to states that invested in innovation and research and development (R&D), then the balance would indeed have little meaning. But technological change is not always predictably linear. Long periods of time may elapse without any discernible changes in the nature of prevailing weapons technologies or in the way wars are fought.[29] Under such conditions, we would actually expect that an accurate depiction of the offense-defense balance would prove useful for some time, until tangible new military applications technological change eventually rendered it obsolete. Simply assuming that states cannot discern these technological trends is overly pessimistic and attributes worst-case thinking to rational actors. Political and military leaders may be expected to be relatively conservative and risk-averse in their assessments, but simply attributing worst-case assessment to leaders and states goes too far.[30] Instead, it is logical to posit that rational leaders will attempt to assess the existing offense-defense balance at a given point (as a critical component of its security environment) in time as well as its likely longevity, and integrate their assessments into military planning while also working to discern how the balance is likely to change in the future. Prudent planning would not only focus on present conditions but also incorporate rational, probabilistic expectations of the future. So Lieber is correct in arguing that leaders don't simply base their policies on an existing balance.[31] Their rationales are far more nuanced and circumspect and clearly forward-looking, anticipating the effects of technological change as well as actions of potential adversaries.[32] Where Lieber is incorrect is in assuming that there is simply no possibility of periods of technological stability, and thus no way for the offense-defense balance to be relevant beyond a discreet point in time.

Clearly, if state leaders perceived the time period between t_3 and t_4 (or even t_5) to be short and quickly elapsing, then cooperation would be almost impossible. A significant shift from defense-dominance to neutrality and/or offense-dominance would rapidly decrease the security of the state by increasing the probability of war, precisely because offensive operations are now feasible, or even preferable. States would be left will little choice but to follow the dictates of the system, exploiting technology to build offensive weapons to deter adversaries or, if necessary, expand to improve their security. While rational, such a decision by one state would likely decrease the security of its neighbors, leading to a security dilemma and an increased probability of war.

However, if military and political leaders were able to expect with some certainty that the system would remain defense-dominant or neutral for some extended period of time, the situation could be quite different. While leaders will still have to confront the possibility that the system will change in the long term, the ability to decrease uncertainty, signal benign motives and thus improve their mutual security may allow them to cooperatively confront that shift to a more offensive world without sparking a security dilemma. In the theoretical best case, security-seeking states would be able to "lock-in" a defense dominant system by negotiating an agreement to forego the development of military applications that would contribute to the execution of offensive operations. As examined in chapter 3 of this volume, this is precisely what many of the proponents of the Preparatory Commission of the World Disarmament Conference were attempting to achieve—albeit unsuccessfully—in the mid-1920s and early 1930s.[33] If applications that provided mobility, firepower, and protection, like tanks, bomber aircraft, and mobile artillery could be prohibited or kept under some system of international control, then the defensive nature of the post-World War I system could be maintained, where applications like static fortifications, artillery, and machine guns provided states with security without threatening others.[34] The inherent challenge to this type of cooperation is the uncertainty that states confront regarding the motives of their neighbors, even over longer periods. So long as a neighbor may be greedy, security-seeking states would be accepting risk and vulnerability to not acquire new offensive applications if the offense-defense balance shifts toward offensive dominance.[35] However, security-seeking states should find opportunities to better understand their neighbors and improve their security through cooperation where such stable technological conditions persist.

Moreover, Lieber's model implies an overly endogenous depiction of technological change that is misleading.[36] While states will rationally seek to develop new weapons, they cannot simply create the next "killer application" because they desire it. The reality of successful innovation is much more complex.[37] Technology can be influenced over time by purposeful state action, but there is also a critical exogenous element to technology that does often present actors with what is close to objective reality, particularly in the short term.[38] The presence of underlying scientific, technical, and industrial capabilities is often necessary for subsequent technological advances.[39] These are important constraints, in large part determining what is possible at any given point in time. Entrepreneurs will have opportunities on the margins, and innovations will diffuse, albeit unevenly, over time, but purposeful actors are not simply presented with unbounded possibilities to harness and exploit technology for their own ends.[40] In the short term potential innovators will remain constrained by underlying scientific-technical structures and processes outside of the direct manipulation of states and will be forced to work within the bounds of prevailing technology as essentially objective reality.[41] Change takes time, and exploiting technological change will similarly be a medium to long-term process almost by definition.

This discussion of the offense-defense balance and technological opportunism illustrates the challenges facing security cooperation and the specific challenge of technological change.[42] It thus provides a useful frame for the introduction of the core of Military Expectations Theory which focuses on the expectations leaders develop regarding the military impact of new technologies, as well as the capacity to effectively understand what other states may develop and deploy over time.[43]

MILITARY EXPECTATIONS THEORY

As great powers constructed organizations to maintain their national security, the development and acquisition of the types of weapons that could improve their security and the surveillance of similar activities by other states emerged as critical tasks.[44] A central premise is that military technology is viewed as a double-edged sword, providing innovators with new applications that could significantly enhance their security.[45] At the same time, prudent security states may be fearful that such capabilities could be used against them by others. To assess their relative positions vis-à-vis potential adversaries, they will develop expectations about the military impact of new technological applications.[46] These expectations will shape how they view the relative military value of potential weapons and focus their scarce resources efficiently.[47] In turn, these expectations will also shape the willingness of leaders to consider restraints or limitations on existing or emerging weapons through arms control.[48] For the purpose of this discussion, a decade to fifteen years is a reasonable period in which state leaders develop expectations.[49] This typically includes processes of research and development (R&D), testing, and production, as well as the time necessary to develop new doctrines and tactics to effectively utilize the new application.[50] Two additional contextual variables, or parameters, will also influence the way in which states develop expectations and their willingness to consider cooperative means of improving their security: the nature of the security environment and the effect of timing of military innovation, which determines whether there is a leading state in a dyad or system.[51]

MILITARY EXPECTATIONS: THE VALUE OF WEAPONS AND THE IMPETUS TO COOPERATE

The military and political leaders of states attempt to mitigate uncertainty in order to develop effective strategies and policies to maintain and improve their security.[52] While it is often a difficult task, leaders will attempt to understand the type of states (whether a fellow security-seeker, greedy, or some combination) they face.[53] Similarly, they must attempt to develop an understanding of the potential opportunities and constraints presented by the security environment, most importantly the implications of technological change.[54] As mentioned above, states cannot simply embrace worst-case thinking, if for no other reason than resource constraints. They simply cannot "have it all" and must invest and devote resources strategically.[55] The expectations that states develop about their neighbors and the capabilities available to defend themselves will drive their strategies and decision-making processes with regard to the relative costs, benefits, and risks of cooperative or competitive policies.[56] Because technological change is an underlying structural force, leaders must focus on the potential impact of new technological applications, typically weapon systems.[57] In turn the expected nature of the weapon system in question will shape the perceived value (or potential threat) to the security of the state.[58] We can think about their expectations relative to new military applications in terms of a set of discrete questions, the first general:

1. How will these applications contribute to state security?

Two specific questions logically influence the answer:

2. What type of threat would these applications pose to state security if acquired and deployed by an adversary?
3. How useful are these applications over existing applications in terms of operational and cost effectiveness?

What is perhaps counterintuitive is that, in large part, the answer to the second question will drive their thinking. Contrary to offensive realist predictions, states do not constantly look for any and all advantage for gain in the security realm. In fact, security-seeking states and their leaders are quite cautious and indeed seem to view the potential threat posed by new applications (the second question) as much more important than the potential benefits (third question). Only when engaged in a clear competitive dynamic—whether an intense security dilemma or a persistent rivalry—will security-seeking states act in a way predicted by offensive realist theory. Under these conditions, where military and political leaders clearly perceive their states to be engaged in an arms race dynamic with an adversary, a security-seeking state will often see little choice but to act competitively and will acquire and deploy new decisive weapons without hesitation.

A *decisive* weapon is one that offers dramatically improved capabilities over comparable existing systems.[59] Whether in terms of firepower, range, mobility, protection, or some combination of those qualities, the acquisition and deployment of these weapons may effectively transform an existing security environment.[60] In offense-defense terms, a decisive weapon (or perhaps a system of discrete weapons) is one capable of shifting a defensive or neutral system to offensive by making offensive operations possible.[61] This would create a dramatic change for states in the system.[62] While rational security-seeking states may prefer the previous status quo where these weapons were not introduced because of their inherent danger, they are nevertheless likely to aggressively pursue the development and deployment of these new applications for fear that another state could beat them to the punch.[63] A state will fear the possibility of a conflict against an adversary that possesses these weapons because their probability of victory will be significantly decreased unless it too possesses them. Moreover, a state that failed to acquire the decisive new weapons would have no guarantee that a neighbor that did would not utilize them. An intense commitment problem emerges.[64] Rational security seekers will be highly concerned about these weapons in the hands of other, but they offer such improvements on existing capabilities that they are highly attractive to military and political leaders, and could (if only possessed by that state) dramatically improve it security through unilateral means. The impulse of security-seeking states may be to find some cooperative means to limit these weapons because of the threat they pose in the hands of others, but for states without these weapons the same commitment problem will lead them to acquire them. In the absence of anything but a benign security environment, cooperation is highly unlikely as states are likely to approach any attempts to restrain these weapons with what are effectively "deadlock" preferences.[65] This may ultimately prove self-defeating over the longer term, if an adversary successfully acquires and deploys these weapons, but the shorter-term considerations are likely to override such concerns.

However, when the capabilities offered by a new weapon system are not perceived as decisive and are likely to provide only *incremental* benefits to existing systems, cooperation becomes possible. This does not mean that these weapons have no perceived value or military utility. In fact, they may constitute central systems in a state's arsenal, but they are expected to provide similar benefits as weapons currently in use.[66] When new weapons are expected to provide only incremental improvements, leaders are less fearful of their acquisition by an adversary so long as they possess them in quantities deemed sufficient to maintain their security vis-à-vis that adversary. The capabilities provided by these weapons cannot overturn the offense-defense balance and thus the conditions for cooperation may arise because the security environment is likely to remain unchanged. Thus leaders may be willing to engage in cooperation to limit or reduce these types of weapons so long as the expected security benefits of cooperation outweigh the potential costs of removing them from the state's arsenal.[67]

It may also be possible that states find that new weapons may have little or no military value, in which case they are likely to be ambivalent about its development or its deployment by an adversary. Conversely states may also be unable to grasp or understand the potential capabilities of a conceptual application. The military impact is highly uncertain, and only a significant period of time will decrease that uncertainty and allow leaders to develop more solid expectations. Cooperation would be difficult in such a situation precisely because leaders do not understand the types of assets that may be available and thus cannot credibly commit to forego them. This is what confronted delegates to the Hague Conferences in 1899 where it was simply too difficult to envision tangible military applications that may emerge from technological changes underway.[68] Similarly, states may find certain projected capabilities to be of particular value, but despite investments in research and development, are unable to solve the technical challenges to make such a weapons systems feasible or cost-effective.

Expectations of the military impact of technological change serves as a useful focus of examination because it can account for changes in leaders' perceptions over time and incorporates the uncertainty inherent in innovation into a model of decision-making. Early in the development of a weapons programs, leaders may develop overly optimistic expectations (or fears) of its capabilities. Over time, as the weapon is developed and deployed and its true capabilities are understood, leaders may adjust their assessment of its perceived value. Similarly, we can think about the underlying technologies as *maturing*, with less uncertainty about potential applications and benefits.[69] Knowing its attributes and limitations, leaders may be more willing to use the weapon as a potential bargaining chip, whereas the previously inflated expectations would have made limiting the weapon undesirable.[70]

The fortuitous conditions for security cooperation do not simply arise when a system's balance is defense dominant or neutral; it is when such a system's balance is perceived as such *and* is expected to remain that way for the foreseeable future. The nature of technological change is perceived as unlikely to provide decisive new capabilities. The military applications that define the period are likely to remain the same or only change in incremental ways, providing limited improvements to existing capabilities in a relevant time period. The knowledge that the incremental nature of improvements will not overturn the existing balance or transform the system provide states with a level of confidence and assurance that they will not be victimized by a breakthrough by another state. As much as offensive realists present the world as one of states looking at all times everywhere to exploit others for their own purposes, it is not rational for states to ignore the systemic opportunities presented by technology, just as it would be irrational to ignore the challenges presented by the system. If the nature of technological change is such that new applications with potential to provide decisive capabilities to innovative states are unlikely to emerge, then states may have the opportunity to cooperate to limit their arsenals, decrease their ability to threaten each other, and improve their mutual security. The problem of systemic technological change remains, however, and states cannot let their guards down. A second critical variable, the effectiveness of existing means of surveillance, remains critical to understanding the conditions for security cooperation.

SURVEILLANCE AND THE CONFIDENCE TO COOPERATE

The expected impact of military technology will drive the willingness of states to consider arms control, but states are most likely to engage in arms control when they are confident that they cannot be victimized by the cheating or defection of their partner. Ultimately, this confidence can only be achieved when a state has the capacity to detect cheating before it is too late. Thus the second technologically influenced independent variable, *the perceived effectiveness of existing modes of surveillance*, is necessary to understanding how states arrive at the decision to cooperate.[71] A critical point to make here, however, is that surveillance must be an existing, indigenous, sovereign capability. For arms control cooperation to be successful, the leaders of a state must be confident that they can adequately assess the capabilities of their potential partner over time, and that any cheating will be detected long before a potential breakout could be achieved. This is indeed a fairly high threshold for surveillance capabilities.[72] The acquisition of intelligence concerning the development programs of potential adversaries must be robust in order to provide leaders that confidence to agree to restrain their own programs.[73] This is not something that can be achieved by an international regime or organization. In an anarchic world, states are unwilling to place their trust in anyone but themselves.[74] Where military technology creates applications that are, by nature, potentially decisive and difficult to independently monitor, states will have little incentive to cooperate in arms control initiatives. Even if the construction of a necessarily intrusive and expansive verification and compliance regime was offered as a solution, states will be exceedingly reluctant to place their security in the hands of a third party if the military impact of that new weapon is expected to be decisive. The potential costs and risks are simply too high.[75]

Instead, states will resign themselves to build the weapons in question even if their eventual acquisition by others will likely make the state less safe and increase the probability of an arms race or war. Such potentially self-defeating behavior would seem irrational, but states have little alternative if they are unable to verify the compliance of partners in a potential arms control agreement. Among security-seeking great powers, where a new application of military technology is expected to play a decisive role in future conflicts, no international regime—no matter how well-constructed—can replace the confidence of possessing those weapons or overcome the fear of fighting an enemy who possesses them without similar capabilities.[76] Only the possession of the indigenous capacity to maintain the level of surveillance necessary to avoid the dangers of defection can provide the setting for a successful arms control agreement. Thus, what has been termed national means of intelligence is typically a second necessary condition in explaining successful arms control outcomes, though the inherent nature of the weapon in question, most importantly its expected capabilities, is also critical.[77]

If a state possesses surveillance capabilities necessary to effectively monitor compliance after the successful negotiation of an arms control agreement, it will be more likely to risk cooperating in the first place precisely because the perceived costs of defection are decreased. What is critical is that the surveillance must not only decrease the risk associated with defection, but also lower the potential costs suffered if an adversary indeed cheats.[78] We would expect, ceteris paribus, that security-seeking states would be highly risk averse to cooperation precisely because they fear defection and the potential costs of defection in the security realm are so high. In some cases, those costs could even entail the survival of the state. How does surveillance lower risks and, more importantly, costs of defection? In the first case, the possession of robust surveillance assets will provide a state with a degree of confidence that it can successfully monitor agreements. Communicating this confidence during negotiations is a prudent tactic, acknowledging that cheating will be detected and serving notice that breaches of good faith will prove self-defeating. Awareness of a state's ability to uncover cheating is likely to deter that behavior in others thus decreasing the risk of defection. More important than minimizing risk, however, is decreasing the costs of cheating, should it occur. Ultimately, this is where robust surveillance will prove decisive. If a state can be confident that it will know with relative certainty that its erstwhile partner is in the process of defecting from an arms control agreement, and can take steps to mitigate the impact of that defection, then it can unilaterally decrease the expected costs of defection with the deployment of effective countermeasures. This may be a fairly high threshold to achieve, and more conservative elements in any state may be unwilling to place the security of the state in potential jeopardy by engaging in cooperation. In reality, satisfying skeptical domestic political and bureaucratic audiences is likely to ensure that the nature of the unilateral surveillance is sufficiently robust that both the risks and costs of defection will be dramatically lowered if indeed an agreement can be forged.[79] This extensive surveillance requirement is unlikely to be achieved by any constructed international regimes, particularly when the weapons in question are perceived to be decisive.[80]

EXPLAINING OUTCOMES: THE INTERACTION OF MILITARY EXPECTATIONS AND PREVAILING MEANS OF SURVEILLANCE

For the purposes of this study, states are able to develop four distinct expectations concerning the military impact of new technologies, and considering the confidence security-seeking states possess in prevailing modes of surveillance, we are presented with a model of predicted outcomes captured by the four-by-two captured in figure 1.2.

The first and most obvious expectation is that a new technology will have no impact. A new application will have no tangible military value. Thus, states have little interest in developing or deploying it, and are ambivalent about whether an adversary does. There is no impetus to seek out cooperation to limit or reduce such weapons, and thus whether one can effectively monitor such a weapon is also irrelevant.[81] Conversely, the fourth column reflects the situation where state military and political leaders may view dramatic changes coming, but they are too far away from actual technical break-

Expectations of Military Impact of New Technology

Prevailing means of Surveillance	**No Impact**	**Incremental Improvement**	**Decisive Improvement**	**High Uncertainty**
High Confidence	**No impetus for Arms Control – States are ambivalent about new weapons**	**Probable** *Qualitative agreements* to limit or reduce certain classes of weapons	**Unlikely** Even with clear picture of adversary's programs, nature of weapon makes cooperation undesirable;	**Not possible to monitor apps that can not be envisioned**
Low Confidence		**Possible** Agreed verification regime can support limits/reductions OR Redundant levels of assets may mitigate uncertainty and reassure;	**Impossible** States cannot credibly commit to forego weapons that are too tempting to possess or too threatening to be possessed by rivals	**Improbable** States don't understand potential impact of new systems; May proscribe behaviors, processes

Figure 1.2 Arms Control: Expectations and Surveillance

throughs or successful developments to grasp the likely utility of new military applications. In this world, it is highly unlikely that any potential weapon that is so conceptually vague and elusive would be amenable to prevailing surveillance capabilities, so the top-left box, where states are relatively confident of such capabilities, essentially drops out. With less confidence, states may be concerned with the implications of future developments of certain technologies, but they are so unsure about the size, shape, and nature of the likely applications that they are unable to construct viable or credible limitations.

Ultimately, the two most relevant sets of conditions emerge when states develop expectations regarding the military impact of new technologies that are either incremental in nature or decisive. As discussed above, in the first case, a state will view new military applications as essentially unchanged from existing weapons, with only marginal improvements in capabilities. This does not mean that these weapons are unimportant or are not perceived as militarily valuable. In fact, they may be the primary systems in a state's arsenal, but their maximum contribution to the execution of central military missions is known and understood with relative certainty.

Where states expect the military impact of new applications to only provide incremental benefits above existing assets, and they possess high confidence in their ability to unilaterally detect cheating by partners, they will support arms control agreements to improve their security through cooperation. Arms control agreements are likely to emerge in the form of reductions or limitations on major assets that are viewed as unlikely to change in the foreseeable future. In this situation, states can feel relatively confident about their security even when engaging in cooperation that limits or reduces numbers of these systems.

Where states are less confident in their ability to monitor the behavior of others and detect cheating, arms control may be possible but it is likely to be difficult, and the agreements that can be forged will be much less robust than when states are confident in their surveillance. Because of the incremental benefits of new systems, states may be more willing to limit or forego them, but without confidence in surveillance they are likely to maintain higher quantities of the weapon in question to maintain required capabilities should cheating occur. Under these conditions, however, the construction of regimes or entities to facilitate monitoring may prove most useful, building upon the cooperation of the initial agreement to further bolster the confidence of the security-seeking states. Various methods such as consultative commissions, inspection regimes and information-sharing processes can contribute to more robust agreements among security seekers.[82] Thus we can understand the potential role and contribution of regimes but only in a circumscribed way in the security realm.

Decisive weapons present states with a far more acute challenge. Because of the capabilities they offer, states see them as both more valuable and more dangerous in the hands of others. Thus, the level of surveillance necessary to provide states with the confidence to cooperate must be perfect. Nothing less would be acceptable to a rational, security-seeking state and its leadership. Where states have little confidence in their surveillance capabilities, any arms control agreement would be impossible. States may indeed be committed to avoiding an arms race in a dangerous new military application. Driven by fears of decisive new weapons, security-seeking states may choose to explore ways to avoid arms races that could increase the probability of conflict. Negotiations may focus on the construction of a robust system of verification and compliance, but ultimately rational, security-seeking states simply are unable to accept a potentially catastrophic vulnerability and place their security, and perhaps survival, in the hands of a regime. Even with relatively high confidence in existing unilateral surveillance capabilities, states are unlikely to be able to cooperate to restrain decisive new weapons because of the perceived vulnerability they would be creating if defection should take place.

TWO KEY PARAMETERS: THE PREVAILING SECURITY ENVIRONMENT AND TIMING OF INNOVATION

The first parameter or antecedent condition is the nature of the security environment in which the states are operating.[83] While it has been mentioned above, it is useful to consider the concept more explicitly. The security environment confronting any given state is naturally going to incorporate political-military factors like polarity (the number of great powers in the system), and the distribution of power, geography, and prevailing military technology. The latter two will contribute to an offense-defense balance, whether that is most applicable between two states or within a system of states. Additional factors like alliances, diplomatic interactions and perceived motivations of states may also play important roles in shaping the nature of a state's security environment. For the purposes of this discussion, we can consider security environments to be benign, uncertain, or threatening. In the former, a state confronts no clear adversary and the system seems relatively defense-dominant. In the latter, a state may face more adversaries in a potentially offensive environment or in a period of rapid technological change. In an uncertain environment, these factors are less clear. States may be unsure of the motives of their neighbors and the military implications of technology may be unclear. Certainly these sit on a continuum, but they broadly capture the political-military realities confronting a state at any given time. Moreover, it is fairly straightforward to see how these can change over time.[84]

The second parameter is related to the timing of military innovation. While technology may diffuse rapidly, in many cases innovation is not simultaneous across a given system of states. Typically one state achieves a “head-start” over its neighbors, though this may be fleeting. This variable simply asks whether the state a leader (or innovative state) or a laggard (confronted by an innovative state)? Given the logic of MET, states in these two different positions may have very different respective views on the relative costs, benefits, and risks associated with security cooperation in the face of technological change. The model assumes that even a lagging great power has the underlying industrial and technical capacity to achieve any breakthrough that an innovator has achieved. However, in reality, the innovative state has a first move advantage. The notion of a leader is clearly more evident and pronounced when discussing decisive new capability. When discussing mature technologies, a leader may possess a quantitative advantage rather than any pronounced qualitative edge. The choices of a leading state will be particularly interesting, given the predictions of both offensive and defensive realism. The former would predict exploiting an advantageous position while the latter would condition the likelihood of cooperation on the underlying environment. Ultimately, including this parameter, like the nature of the security environment, sharpens our understanding of the conditions under which security cooperation may take place in an anarchic international system and exposes the limitation of existing theoretical approaches.

IMPLICATIONS

MET explains how security cooperation may prove difficult, if not impossible to achieve even where a security environment may be relatively benign. In general, it would be irrational for states to engage in cooperation if there was a risk that they would suffer significant damage, perhaps even their survival, in the event of cheating. While some scholars have made the argument that states must actually accept a significant level of vulnerability to provide signals costly enough to transmit benign intentions, a more logical path to cooperation is to reassure other security-seeking states by foregoing an opportunity that would make that decrease that state’s security through cooperation.[85] MET may help us better understand these conditions.

Conversely, even in a relatively threatening security environment, where two or more security-seeking states may be engaged in a relatively intense security dilemma, the nature of military technology may offer opportunities to reverse spirals of hostility and dampen escalation dynamic through signals of benign intent and, ultimately, arms control cooperation. A successful agreement may even decrease the probability of war. By controlling for the security environment and considering the influence of being a leader or lagging state in terms of the development and deployment of armaments, MET can generate testable hypotheses that are distinct from existing theoretical approaches and more effectively explain the likelihood of cooperation in the security realm.

This approach raises serious questions for offensive realist predictions that view security cooperation as impossible.[86] While the risks inherent in consistently engaging in hardline policies seem clear, this model explains and predicts when states should choose to cooperate or engage in competition based on the underlying technologically driven influences of the security environment.[87] Similarly, while more in line with the underlying logic of defensive realism, the model shows how even rational, security-seeking states may engage in suboptimal or potentially self-defeating behavior because they are unable to forego the potential security benefits provided by decisive new weapons, even if an adversary is likely to acquire them, and even if that acquisition would ultimately leave the first state less secure.[88] The pressures placed on security-seeking states by technological breakthroughs are likely to drive decision-making. Decisive new applications will simply be too tempting to forego or too frightening to ignore in the hands of an adversary. A short "head start" for one state may seem to be the only potential advantage to be exploited, even if ultimately an arms race that one could lose is the likely outcome.

MET also presents a significant challenge to the neoliberal or institutionalist literature in the security realm. It improves our understanding of cooperation by illustrating where regimes are most likely to have an impact on the effective implementation of negotiated settlements, but it also clearly underscores the underlying limitations of an institutional approach. Even when successful, there is little independent role for institutions in explaining security cooperation; the outcomes are determined by structural factors, primarily technological in nature. Ultimately, for rational, security-seeking states, technological change can make cooperation exceedingly difficult even under relatively favorable political conditions. More importantly, the absence of rapid technological change and the emergence of decisive new applications can allow states to cooperate and improve their security even under less favorable political conditions. The potential benefits of security cooperation will be discussed in the next section.

THE BENEFITS OF SECURITY COOPERATION

Often overlooked in discussions on security cooperation is the fact that successful arms control agreements can provide real, tangible security benefits to participating states. Without the prospect of an improvement on their relative security, states would have little interest in pursuing cooperative approaches over unilateral or even competitive means.[89] As discussed, security-seeking states in the model are assumed to understand the potential downsides of hardline policies, and the self-defeating aspects of unilateral (and particularly competitive) strategies. Cooperation is risky, but so is competition.[90] They understand that they could potentially lose an arms race to a competitor making them less likely to prevail in a subsequent war made more likely by the emergence of an arms race. Nonetheless, without some prospect of actual benefits deriving from potential cooperation, it would seem unlikely that states would accept the expected risks and probable costs to their security should cooperation fail. Assessing the potential costs, benefits, and risks of cooperative policies is a difficult task for any state. Contrasting the costs, benefits, and risks of cooperation against unilateral or competitive policies will be even more challenging in an uncertain world.[91] However, if structural factors are able to provide states with opportunities to negotiate and conclude arms control agreements, the security benefits can be significant. There are four related but distinct contributions that arms control can provide to states. The first two contributions, enhancing arms race stability and supporting crisis stability are both related to decreasing uncertainties that could lead to inefficient or suboptimal outcomes arising from unilateral policies and actions. Signaling intentions through arms control initiatives can support a state's strategy of reassurance of potential adversaries, or learning more about their intentions. The final contribution, decreasing the probability of war through the dampening of security dilemma dynamics, is obviously the most important and makes the greatest contribution to a state's security but it is also the most difficult to achieve.

Arms race stability refers to a condition in which two sides compete in a relatively normalized and transparent way.[92] Instabilities arise where states are concerned with a dramatic new technological application providing a decisive advantage to their adversaries.[93] In such situations, leaders may view potential windows of vulnerability when an adversary achieves a breakthrough.[94] This may in turn spur preventive motives in the laggard state to strike before falling too far behind in either qualitative or quantitative terms.[95] Conversely, the leading state, achieving a breakthrough may have motives to exploit its advantages over a rival by launching a preventive strike. Arms control can contribute to the security of both states by removing potential areas of conflict that could provide decisive new capabilities if competition persisted unabated.[96] Typically, the applications in question may be exceedingly difficult in technical terms to achieve without massive investments over a long term. Agreement to forego efforts to develop and deploy such weapons can regularize competition and contribute to stable, shared expectations even in the presence of a persistent rivalry. At the very least, such cooperation can avoid wasting resources that would otherwise be used productively.

States can also enhance *crisis stability* by cooperatively restraining certain types of forces or weapons that may provide one or both sides with an incentive to strike first in the event of a diplomatic crisis.[97] While this became a much more pressing need during the Cold War in the nuclear context, it is also possible to consider how such measures could be designed to support stability in the conventional context. Deployment schedules and force postures can be purposefully developed to avoid the incentive for launching a first-strike in the event of a crisis. The inherent difficulty in the conventional context is the overriding concern of succeeding in a military conflict (effective defense). The deterrent effect of nuclear weapons, based on some recognition of assured destruction, that arose during the Cold War is absent in the conventional context. Nonetheless, security-seeking states should be able to develop confidence building measures that would constrain, rather than contribute to first-strike incentives during a crisis, removing the pressures on states and leaders to resort to the use of military force.

Enhancing arms race and crisis stability are specific examples of what could more broadly be considered *signaling benign intentions* to others.[98] States may utilize their military policies to attempt to decrease the concerns of neighbors by exhibiting restraint and willingness to forego potential advantages.[99] As the literatures on signaling and defensive realism reflect, signals may not always be understood. Moreover, structural constraints on behavior may make it impossible for states to respond to ostensibly benign signals even if they objectively sought to do so.[100] For example, if a security environment was offense dominant, it may simply be too risky to initiate or reciprocate a signal of benign intent because of the necessary level of vulnerability the state would have to accept to make the signal costly. However, if the security environment (the geography and prevailing military technologies) combined to allow states to signal benign intentions through the limitation on certain types of quantities of weapons or personnel, then arms control may allow states to decrease uncertainty and mistrust.[101] These can be small steps at first, but after various iterations, states may succeed in dramatically altering their perceptions of each other through repeated signaling utilizing arms control.

Finally, the most important form of security benefit derived from arms control cooperation is to *decrease the probability of conflict*.[102] Certainly, maintaining arms race stability and decreasing first strike motives would generally contribute to decreasing the probability of conflict, but they do so indirectly, through the mitigation of uncertainties that could drive states to act for fear of being exploited. In a much more straightforward way, states can decrease the probability of war by utilizing arms control to limit or reduce the threat they pose to others and thus remove the sources of insecurity.[103] Where rational security-seeking states are locked in a security dilemma, unintentionally threatening each other with unilateral policies, and are unable to discern the type of actor they are interacting with, arms control may provide an opportunity to escape the spiral of mistrust and decrease the probability of conflict. Cooperative agreement to limit certain types or quantities of weapons that are most likely to be utilized in a future war will allow states to maintain their own security without threatening others. This is obviously a critically important, though often unusual, outcome.

BARGAINING ON THE CURVE: TESTABLE HYPOTHESES AND ALTERNATIVES

As MET explains, because technology is dynamic—it is always changing, but not always at the same rate of speed—leaders are driven to attempt to understand the implications of that change for their security. Military and political leaders will develop expectations about the military applications of new and emerging technologies in order to inform decisions about the dedication of scare resources to provide for the security and survival of their state moving forward. Ultimately, the success of any agreement is likely to hinge on the confidence leaders have in existing means of surveillance to monitor compliance and avoid defection.

MET GENERAL PROPOSITIONS

Building upon this discussion of military technology and its influence on the decision-making and policies of security-seeking states, we can construct several clear, testable hypotheses that will be assessed against the historical cases of arms development and acquisition, arms races, and arms control cooperation. Three general propositions include:

$\mathbf{H}_{\mathbf{MET1}}$ When security-seeking states view new weapons as decisive, arms control is impossible.

$\mathbf{H}_{\mathbf{MET2}}$ When security-seeking states view new weapons as incremental improvements upon existing systems, arms control is possible.

$\mathbf{H}_{\mathbf{MET3}}$ When security seeking states view new weapons as incremental improvements, and prevailing means of surveillance are perceived as effective in detecting potential violations, arms control is likely.

When incorporating the parameters of timing of innovation and the nature of the prevailing security environment in which innovations take place to the model, several interesting hypotheses that derive from the logic of MET concerning the more specific security strategies of states are apparent.

LEADERS AND LAGGARDS BARGAINING ON THE CURVE

H_{Lead1} A leading (innovative) security-seeking state may be willing to bargain on the limitation of a decisive weapon in a relatively benign security environment.

Cognizant that it (1) possesses present advantages in existing weapons, (2) could lose an arms race in the decisive new weapon, (3) may perceive the new weapons a specifically threatening to its own security in the hands of an adversary—a security-seeking state may desire to avoid a race in the decisive new weapons system if the relationship with the adversary is ambiguous.

H_{Lead2} A leading security-seeking state will be unlikely to consider bargaining on a decisive weapon in a threatening or uncertain security environment.

The nature of the security environment in which an innovation takes place thus conditions leaders of security-seeking states about the potential systemic impact of their breakthrough. Security-seeking states may prefer the previous status quo, viewing the weapon as particularly threatening to their own security and prefer a world without them. However, the offer is unlikely to be reciprocated and without guarantees about effective compliance and monitoring of agreements, arms control would be unlikely.

H_{Lead3} A leading security-seeking state will be willing to bargain on the limitation of existing weapons or new systems that are deemed to be of incremental military value.

While possession of sufficient quantities of such a system may provide requisite security objectives, the actual success of cooperation will depend on surveillance capabilities.

For a laggard state, the situation is different. So long as it possesses the underlying technical and industrial capabilities to maintain great power status, the achievement of a truly decisive military innovation or technological breakthrough could be expected to create an acute sense of vulnerability in the laggard state. The only means of rectifying this highly threatening situation is to attain the breakthrough application. Thus any cooperative initiatives to prohibit or restrain these decisive weapons would lock-in the laggard state to a position of vulnerability which would be unacceptable to any rational security-seeking state.

H_{Lag1} A laggard security-seeking state is highly unlikely to accept limitation on a decisive weapon. Agreed limitations cannot replace the know-how and capacity to achieve the weapon.

However, if the prospects of arms limitation or reduction involving a leading state arise, a laggard state will be willing to engage in negotiations is the ultimate agreement is perceived as improving its security. Nonetheless, because it is in an inferior position, it is likely to bargain hard for concessions and quantitative levels that improve its position vis-à-vis the leader.

H_{Lag2} A laggard security-seeking state is likely to accept limitations on existing weapons or new systems that are deemed to be of incremental value if they possess these systems in sufficient quantities.

Provided the agreement includes these preferences, the laggard state is likely to cooperate.

In the absence of a clear leader, or in a truly multipolar environment, great powers may achieve breakthrough more or less simultaneously. Given the presence of several great powers, we can assume that the prevailing security environment is likely to be relatively uncertain or perhaps even threatening. In these cases, new weapons are likely to exacerbate underlying difference between states and decrease stability.

H_{Sim} Relatively simultaneous development and deployment of decisive new weapons is likely to increase perceived threats and decrease the probability of cooperation among states in a given system.

ALTERNATIVE SYSTEMIC HYPOTHESES

Taking the hypotheses derived from state's expectations of the impact of military technology, it is useful to develop several robust alternative hypotheses to test against the empirical record in the case studies ahead. To begin with offensive realism, the theory really seems to have little to add to our discussion. It is frankly hard to see how offensive realism would comment on arms control. It may perhaps, in its basic forms provide a null hypothesis:

AH_{OFF1} States will never engage in "meaningful" arms control.

Given the uncertainty of the political environment, statesmen in an offensive realist world will rely on worst-case analysis with regard to the acquisition and deployment of new technologies. Even in a situation where technology is static or stable, they will be loath to engage in cooperative measures that would limit the fielding of capabilities in the future. Thus, it is really difficult to see statesmen agreeing to cooperative initiatives other than when such initiatives are deemed as "cosmetic" and of little real impact to their ability to acquire and deploy military assets in the present or future. This would lead to a more cynical alternative hypothesis.

AH_{OFF2} Statesmen will only agree to short-term measures in a benign security environment. They will not bind themselves to agreements that are perceived to constrain their abilities to effectively provide for the security of their states.

As mentioned above, perhaps, in benign security environment, which is unlikely in an offensive realist world, engagement in such cooperative endeavors is perceived as useful, so as not to signal greedy motives and unnecessarily anger other states. But even this line of thinking seems to preclude a much more circumspect offensive realist leader than the ones we often see depicted in much of the literature.

Defensive realist approaches view cooperation as a potentially viable means improving a state's security and avoid unnecessary arms races or conflict between security-seeking states. But since successful arms control has been relatively scarce, we must ask why we do not see more as prescribed by defensive realism. As discussed above, the expectations of the impact of military technology refine defensive realist arguments, and further delineate the conditions under which we may expect cooperation to be possible.

> **AH_{DEF1}** When a security environment is relatively benign, statesmen are more willing to engage in cooperative arms control initiatives. Conversely, when a political environment is viewed as tenuous or dangerous, statesmen will be unwilling to engage in cooperative activities such as arms control.

From the perspective of MET, the problem with this hypothesis, which reflects the flaw in many defensive realist analyses, is that it is relatively static. Introducing a dynamic notion of technology (to better test against the hypotheses offered above) and to incorporate the political outcomes that should follow from defensive realist predictions, the following alternative hypothesis improves upon the previous one:

> **AH_{DEF2}** When a security environment is relatively benign, statesmen should take serious steps to constrain technological innovation that may overturn the existing system.

Why focus on the benign environment? When there is tension, it is often too much to ask for states to engage in cooperation if the tradeoffs are relatively high. Such a situation exists in the "offense-dominant/able to distinguish" category. It is true that arms control would be most fruitful in this environment, but the costs of absorbing a first strike are such that the risks of defection are simply too great. But when states are in a defense dominant system, we would expect them to cooperatively engage in efforts to maintain that system through cooperation that would most likely be focused on limiting offensive weapons. This is the logical policy of a defensive realist theory, and it seems rational and within the realm of the possible.

Turning briefly to institutions, given the difficult nature of cooperation in the security realm, we would expect that institutions provide the greatest amount of utility to security-seeking states, and thus exert the most independent power, where they can reassure states against the risks of defection and (ideally) help to minimize the costs should defection occur. Thus a relatively straightforward alternative hypothesis is the following:

> **AH**$_{NLI1}$ International institutions improve the likelihood of cooperative agreements.

Institutions facilitate arms control by (1) providing venues for discussion and dialogue; (2) linking arms control to other issues of importance to states, making the development of sustainable agreements more likely; and (3) providing services to states in the form of monitoring, technical expertise, and enforcement.[104] Any combination of these three mechanisms may indeed help states create negotiated settlements that improve their security. We may particularly expect that the third mechanism would be of special interest in the realm of arms control. Verification and monitoring of states to maintain compliance with previously agreed limitations will be essential if an agreement is to be sustained.

ALTERNATIVE HYPOTHESES: DOMESTIC FACTORS AND SECURITY COOPERATION

While this chapter has focused primarily on structural explanations for security cooperation, other scholars have posited the importance in unit-level, domestic factors that may play a larger role in explaining state behavior than systemic imperatives. Looking inside the "black-box" of the state, they argue that factors such as regime type, economic system, and ideology (among others) often determine a state's behavior, or at least circumscribe its ability to respond to structural imperatives.[105] Only in the presence of certain domestic configurations can states truly act like the rational unitary actors assumed in most structural theories, and domestic political and economic constraints will ultimately determine policies.[106] Domestic factors may best explain why states exhibit pathological behaviors in the face of what should be seen as structural constraints engaging in over-expansion or aggression.[107] Conversely, in a system inhabited by states that value interdependence, we would expect to see high levels of trade and commerce.[108] Moreover, as more states become liberal and democratic, we would expect to see a zone of relative peace between states with like institutions, shared values, and ideologies.[109]

These domestic approaches can provide interesting analyses and explain a great deal about how a state came to develop a certain strategy or policy.[110] However, in attempting to develop more generally hypotheses, with regard to the research, development and acquisition of arms, one domestic factor, economic concerns, could play several (some countervailing) roles.[111] We can logically argue that states facing fiscal constraints or economic depressions would seek arms control to avoid arms races. But conversely, similar states facing similar constraints could see investment in defense as a way to stimulate the economy, or may see the subsidizing of an arms industry as an important national objective, perhaps driven by parochial military interests. It is thus unclear, without investigation, in which direction "economic hardship" would influence an arms buildup: supporting or restraining. For the purposes of this study, two distinct alternative hypotheses derived from domestic-level approaches are particularly germane and have often been discussed as influencing international cooperation. The first relates to resource constraints.

> **AH_{DOM1}** Facing significant domestic budgetary constraints, leaders will seek out and respond favorably to arms control initiatives to avoid costly military expenditures.

This goes well beyond the assumption of resource constraints incorporated in MET. It argues that domestic imperatives are really driving the foreign policies of the states. To alleviate difficult fiscal pressures at home, leaders will seek to avoid paying for costly weapons programs. Therefore, the ultimate success of an arms control agreement owes more to the effective coordination of states' domestically driven economic preferences than to the achievement of perceived security benefits through cooperation.

The second hypothesis relates to worldviews and argues that successful arms control results from states acknowledging shared values or ideas about the nature of the international system. In a very basic way, this could be understood as analogous to logic the liberal or democratic peace argument. However, for the purposes of this study, it can also include major strategic doctrines or view or prevailing views on the nature of security. A liberal peace argument would have little relevance in two Cold War cases, nor would it be particularly applicable in the context of the ideological diversity of the 1920s and 1930s.

> **AH_{DOM2}** States will agree to arms control when they share similar views about the nature of international security and appropriate means of providing for national security.

So for example, if the leaders of the Soviet Union came to agree that Mutually Assured Destruction was indeed the appropriate way of viewing the arms race and the superpower competition, then this, rather than MET, would explain the cooperation found in the SALT case. Similarly, shared views of naval warfare would provide the explanation for success at the Washington Naval Conference.[112] More traditional liberal peace predictions can also be tested (among the United States, Great Britain, and France) as well, but it is useful to broaden this alternative hypothesis to apply to all of the cases.

CONCLUSION

In the following chapters, the hypothesis and the model developed above will be put to the test. We will examine four periods of arms control negotiations to attempt to measure the influence of technological expectations against the alternative hypotheses. Utilizing documentary evidence, as well as other primary and secondary sources, it should be possible to reconstruct the expectations of relevant political and military leaders. In fact, the cases will be constructed to present just such an *ex ante* presentation of the military expectations of technological change. Each case study will first present a depiction of the security environment prior to the cooperative engagement in arms control negotiations. An analysis of the expectations and perceptions of the various states' military and political leaders will follow. A review of the actual negotiations in each case will help us understand how expectations may be a necessary but not sufficient condition for explaining success in arms control. Finally, the conclusions of the case studies will recapitulate the performance of MET in each case and evaluate the arguments of alternative hypotheses.

The cases consist of two interwar, multilateral, great power negotiations over conventional weapons, and two Cold War bilateral superpower negotiations focused on nuclear weapons. Thus there is significant diversity within the cases on a number of interesting dimensions including the prevailing security environments, the number of participants, and the technologies and applications under consideration. At the same time, the four cases share one crucial similarity. In each case, the states involved were attempting to proactively address the potential problem of competitive armament and arms racing with the ultimate goal of dampening conflict and decreasing the probability of war through cooperation. With great foresight, these leaders were attempting to avert disaster through diplomacy and arms control. They were making assessments concerning the likelihood of conflict in the future, not unlike their military counterparts, and saw the potential for conflict as increasing. Through the limitation of armaments, specifically certain classes of armaments, or qualitative arms control, they sought to improve the security of the participating states through cooperation. In fact, what they were grappling with were precisely the dynamic forces of technological change over time in the context of an uncertain, anarchic international system. Where they were successful, military expectations allowed for perceptions of stability that allowed for credible commitments, while in the cases of failure, the decisive and dynamic perceptions of technology simply made commitment impossible.

NOTES

1. John J. Mearsheimer, "The False Promise of International Institutions," *International Security* 19, no. 3 (1994/95).

2. Dale C. Copeland, "Theory and History in the Study of Major War," *Security Studies* 10, no. 4 (2001): 213–14.

3. Charles L. Glaser, "Realists as Optimists: Cooperation as Self-Help," *International Security* 19, no. 3 (1994/95); Robert Jervis, "Cooperation under the Security Dilemma," *World Politics* 30, no. 2 (1978); Stephen Van Evera, "Offense, Defense and the Causes of War," *International Security* 22, no. 4 (1998).

4. Robert Axelrod and Robert O. Keohane, "Achieving Cooperation under Anarchy: Strategies and Institutions," *World Politics* 38, no. 1 (1985); Helga Haftendorn, Robert O. Keohane, and Celeste A. Wallander, eds., *Imperfect Unions: Security Institutions over Time and Space* (Oxford: Oxford University Press, 1999); Robert O. Keohane, "The Demand for International Regimes," *International Organization* 36, no. 2 (1982); Robert O. Keohane and Lisa L. Martin, "The Promise of Institutionalist Theory," *International Security* 20, no. 1 (1995).

5. Haftendorn, Keohane, and Wallander, *Imperfect Unions*.

6. Charles Lipson, "International Cooperation in Economic and Security Affairs," *World Politics* 37, no. 1(1984): 1–23.

7. John J. Mearsheimer, *The Tragedy of Great Power Politics* (New York: W. W. Norton, 2001); Robert Gilpin, *War and Change in World Politics* (Princeton, NJ: Princeton University Press, 1981); Dale C. Copeland, *The Origins of Major War* (Ithaca, NY: Cornell University Press, 1997).

8. Mearsheimer, "The False Promise of International Institutions," 11–12.

9. Ibid., 19–26.

10. Charles L. Glaser, "The Security Dilemma Revisted," *World Politics* 50, no. 1 (1997): 171–201; Glaser, "Realists as Optimists," 71–72.

11. Copeland, "Theory and History in the Study of Major War."

12. Charles L. Glaser and Chaim Kaufmann, "What Is the Offense-Defense Balance and Can We Measure It?," *International Security* 22, no. 4 (1998).

13. Jervis, "Cooperation under the Security Dilemma," 188.

14. Ibid., 186–214.

15. Stephen Van Evera, *Causes of War: Power and the Roots of Conflict* (Ithaca, NY: Cornell University Press, 1999), 7–11.

16. Jonathan Shimshoni, "Technology, Military Advantage, and World War I: A Case for Military Entrepreneurship," *International Security* 15, no. 3 (1990).

17. Charles L. Glaser, "Political Consequences of Military Strategy: Expanding and Refining the Spiral and Deterrence Models," *World Politics* 44, no. 4 (1992).

18. Evan Braden Montgomery, "Breaking out of the Security Dilemma: Realism, Reassurance and the Problem of Uncertainty," *International Security* 31, no. 2 (2006).

19. Andrew Kydd, "Sheep in Sheep's Clothing: Why Security Seekers Do Not Fight Each Other," *Security Studies* 7, no. 1 (Autumn 1997): 114–55.

20. Glaser, "Realists as Optimists: Cooperation as Self-Help," 70.

21. Van Evera, *Causes of War*, 35–72.

22. On other criticisms, see Karen Ruth Adams, "Attack and Conquer? International Anarchy and the Offense-Defense-Deterrence Balance," *International Security* 28, no. 3 (2003/2004); Richard K. Betts, "Must War Find a Way?," *International Security* 24, no. 2 (1999); James W. Davis Jr. et al., "Taking Offense at Offense-Defense Theory," *International Security* 23, no. 3 (1998).

23. Dale C. Copeland, "Rationalist Theories of International Politics and the Problem of the Future," *Security Studies* 20, no. 3 (2011); Charles L. Glaser, *Rational Theory of International Politics: The Logic of Competition and Cooperation* (Princeton, NJ: Princeton University Press, 2010).

24. Barry Buzan and Eric Herring, *The Arms Dynamic in World Politics* (Boulder, CO: Lynne Rienner, 1998), 19–28; Herbert F. York, *Race to Oblivion: A Participant's View of the Arms Race* (New York: Clarion, 1970), 153–56.

25. Buzan and Herring, *The Arms Dynamic in World Politica*, 12.
26. Keir A. Lieber, *War and the Engineers: The Primacy of Politics over Technology* (Ithaca, NY: Cornell University Press, 2005).
27. Ibid., 4–7.
28. Ibid., 4.
29. Jack S. Levy, "The Offensive/Defensive Balance of Military Technology: A Theoretical and Historical Analysis," *International Studies Quarterly* 28, no. 2 (1984).
30. Glaser, *Rational Theory of International Politics*, 70.
31. Lieber, *War and the Engineers: The Primacy of Politics over Technology*, 4.
32. Stephen J. Majeski, "Expectations and Arms Races," *American Journal of Political Science* 29, no. 2 (1985).
33. B. J. C. McKercher, "Of Horns and Teeth: The Preparatory Commission and the World Disarmament Conference, 1926–1934," in *Arms Limitation and Disarmament: Restraints on War, 1899–1939*, ed. B. J. C. McKercher (Westport, CT: Praeger, 1992); Philip Noel-Baker, *The First World Disarmament Conference, 1932–1933 and Why It Failed* (Oxford: Pergamon Press, 1979); John W. Wheeler-Bennett, *The Pipe Dream of Peace: The Story of the Collapse of Disarmamen*t (New York: William Morrow, 1935).
34. This assumes that actors could indeed differentiate between weapons that would prove most effective for offensive operations and maintaining defenses. For an alternative view, see John J. Mearsheimer, *Conventional Deterrence* (Ithaca, NY: Cornell University Press, 1983).
35. Glaser, *Rational Theory of International Politics*.
36. Lieber, *War and the Engineers: The Primacy of Politics over Technology*, 5.
37. Geoffrey L. Herrera, *Technology and International Transformation: The Railroad, the Atom Bomb, and the Politics of Technological Change* (Albany: State University of New York Press, 2006); Barry Watts and Williamson Murray, "Military Innovation in Peacetime," in *Military Innovation in the Interwar Period*, ed. Williamson Murray and Allan R. Millett (Cambridge: Cambridge University Press, 1996).
38. Geoffrey L. Herrera, "The Mobility of Power: Technology, Diffusion and International Change" (PhD dissertation, Princeton University, 1995), 15.
39. Buzan and Herring, *The Arms Dynamic in World Politics*; William H. McNeill, *The Pursuit of Power: Technology, Armed Force, and Society since A.D. 1000* (Chicago: University of Chicago Press, 1984); Martin Van Creveld, *Technology and War: From 2000 B.C. to the Present* (New York: Free Press, 1989).
40. Herrera, "The Mobility of Power," 15.
41. Ibid.
42. A plausible first-cut theory could focus on the prospects of cooperation during periods of rapid technological change versus periods of relative stability. The difficulty here would be ex-post coding of different time periods and establishing successes or failures of arms control cooperation. There are seemingly two problems with this approach. First, as the cases in this study indicate, leaders must bargain over limitations or reductions of discreet, tangible assets. Even in successful cases, like Washington Naval or SALT I, there is variation on success in failure within the cases: some weapons were limited and some were not. Second, and more generally, the military implications of technology are going to be filtered through the leaderships (and the military and political leaders) of states and thus the focus of this paper (rightly) is the perceptions of those leaders. Simply looking back and coding time periods as "rapid" or "stable" and attempting to measure cooperative or competitive pressures on states is likely to obscure rather than clarify the perceptions and intentions of leaders at the time.
43. MET is still considered a structural realist theory because it assumes that states are acting "as if" they were rationally responding to the constraints and opportunities created by the system, in this case focusing primarily on the influence of technological change. Though the level of analysis utilized may be focused within states (even on individual leaders), it affirms structural predictions.
44. This study assumes that states possess the capacities to engage in developing new military applications and is thus applicable to states from the late 1800s to the present. Kenneth M. Waltz, *Theory of International Politics* (New York: McGraw-Hill, 1979), 76.
45. Mearsheimer, *The Tragedy of Great Power Politics*, 166.

46. Stephen J. Majeski, "Technological Innovation and Cooperation in Arms Races," *International Studies Quarterly* 30, no. 2 (1986).

47. Ibid., 175–91.

48. Majeski, "Expectations and Arms Races," 222.

49. Marek Thee, *Military Technology Military Strategy and the Arms Race* (New York: St. Martin's Press, 1986), 16.

50. Millett, "Patterns of Military Innovation in the Interwar Period," 359; Stephen Peter Rosen, *Winning the Next War: Innovation and the Modern Military* (Ithaca, NY: Cornell University Press, 1991), 19–21.

51. These can also be considered antecedent conditions in methodological terms. These can be understood as "a phenomenon whose presence activates or magnifies the action of a causal law or hypotheses." Stephen Van Evera, *Guide to Methods for Students of Political Science* (Ithaca, NY: Cornell University Press, 1997), 9.

52. Barry Posen, *The Sources of Military Doctrine: France Britain and Germany between the World Wars* (Ithaca, NY: Cornell University Press, 1984).

53. Glaser, *Rational Theory of International Politics*; Andrew Kydd, "Trust, Reassurance, and Cooperation," *International Organization* 54, no. 2 (2000).

54. Posen, *The Sources of Military Doctrine*.

55. The concept of resource constraints was central to early analyses of arms racing. See Anatol Rapoport, "Lewis F. Richardson's Mathematical Theory of War," *Conflict Resolution* 1, no. 3 (1957); Lewis F. Richardson, *Arms and Insecurity* (Pittsburgh: Homewood, 1960).

56. Glaser, "Political Consequences of Military Strategy: Expanding and Refining the Spiral and Deterrence Models"; Glaser, "Realists as Optimists"; Posen, *The Sources of Military Doctrine*.

57. States cannot credibly agree to restrain technologies or technological change. Thus, they must focus on tangible applications, which can be actually be limited or reduced. See Ted Greenwood, "Why Military Technology Is Difficult to Restrain," *Science, Technology and Human Values* 15, no. 4 (1990).

58. Like its realist cousins, MET assumes that the system is inherently anarchic and states must provide for their own survival. These states are rational and strategic and focused on maximizing their security. In contrast to offensive realism, they see the downsides of constantly employing hardline policies, particularly in the presence of other security-seeking states. However, they are also aware that greedy states with revisionist goals that may value the benefits of expansion over the costs of defeat may existing in the system and thus are constantly attempting to understand both systemic realities (distribution of power, offense-defense balance) and the nature of their neighbors (security-seeking, greedy, etc.). These factors constitute a state's security environment.

59. James R. Kurth, "A Widening Gyre: The Logic of American Weapons Procurement," in *American Foreign Policy Theoretical Essays*, ed. G. John Ikenberry (New York: HarperCollins, 1996); Andrew Kydd, "Arms Races and Arms Control: Modeling the Hawk Perspective," *American Journal of Political Science* 44, no. 2 (2000).

60. Williamson Murray, "Innovation: Past and Future," in *Military Innovation in the Interwar Period* (Cambridge: Cambridge University Press, 1996).

61. In a defense-dominant or "deterrence dominant" or neutral system, offensive weapons may be viewed as potential bargaining chips, though such weapons would be viewed with ambivalence in the former two situations. Thus the focus here on "decisive" weapons thus attempts to broaden the potential applicability of the concept under different contexts. In the nuclear era, a relatively effective strategic defense system would also likely be a "decisive" system because of its impact on the deterrent balance created by assured retaliation capabilities. Herbert F. York, "ABM, MIRV, and the Arms Race," *Science* 169, no. 3942 (1970).

62. The situation presented follows from Kydd's discussion of "military value of arms" as "high." See Andrew Kydd, "Game Theory and the Spiral Model," *World Politics* 49, no. 3 (1997).

63. Harry G. Gelber, "Technical Innovation and Arms Control," *World Politics* 26, no. 4 (1974); Sanford Lackoff and W. Erik Bruvold, "Controlling the Qualitative Arms Race: The Primacy of Politics," *Science, Technology and Human Values* 15, no. 4 (1990).

64. James D. Fearon, "Rationalist Explanations for War," *International Organization* 49, no. 3 (1995); Robert Powell, "Bargaining Theory and International Conflict," *Annual Review of Political Science*, no. 5 (2002); Robert Powell, "War as a Commitment Problem," *International Organization* 60 (2006).

65. George W. Downs, David M. Rocke, and Randolph M. Siverson, "Arms Races and Cooperation," *World Politics* 38, no. 1 (1985).

66. York, *Race to Oblivion*, 160–61.

67. Glaser, "Realists as Optimists," 65–66.

68. Calvin De Armond Davis, *The United States and the First Hague Peace Conference* (Ithaca, NY: Cornell University Press, 1962); James Brown Scott, ed., *The Hague Peace Conferences of 1899 and 1907*, vol. 2 (Baltimore: Johns Hopkins Press, 1900).

69. I am indebted to Robert Reardon for this point.

70. York, *Race to Oblivin*, 70–105, 47–53.

71. Glaser, *Rational Theory of International Politics*.

72. Herbert Scoville, "A Leap Forward in Verification," in *Salt: The Moscow Agreements and Beyond*, ed. Mason Willrich and John B. Rhinelander (New York: Free Press, 1974).

73. This is not to say that surveillance is inherently stabilizing or cooperation-inducing. The ability to acquire and interpret critical information about an adversary's military capabilities in a timely manner can certainly be perceived of—and utilized as—a weapon. At the operational or tactical levels, improved situational awareness will aid military leaders in developing effective operational plans and tactical adjustments, thus making a potentially decisive contribution to successful battlefield outcomes. Good intelligence is a precious commodity indeed. The applications that contribute to a robust surveillance capability, like most technological applications, are tools that can be utilized in different ways. Nonetheless, for a rational, security-seeking state attempting to improve its probability of survival in an anarchic system without threatening other states, surveillance capabilities—and the confidence that those capabilities are robust—will prove extremely useful in avoiding potential suboptimal outcomes. Robust surveillance capabilities may allow states to decrease the uncertainty inherent in the international system, and provide opportunities to explore cooperation with some degree of confidence that defection will be detected before damage can be done. See Buzan and Herring, *The Arms Dynamic in World Politics*; Van Creveld, *Technology and War*.

74. Bernard T. Feld, "Inspection Techniques of Arms Control," in *Arms Control, Disarmament, and National Security*, ed. Donald G. Brennan (New York: Braziller, 1961).

75. Robert Jervis, "Realism, Neoliberalism, and Cooperation: Understanding the Debate," *International Security* 24, no. 1 (1999): 51–52; Charles Lipson, "International Cooperation in Economic and Security Affairs," 19; Mearsheimer, "The False Promise of International Institutions."

76. David W. Kearn Jr., "The Baruch Plan and the Quest for Atomic Disarmament," *Diplomacy and Statecraft* 21, no. 1 (2010).

77. Scoville, "A Leap Forward in Verification," 160–67.

78. In the most favorable circumstances, states will possess a *surplus* of security, which all but ensures that they cannot be victimized by the transgressions of an adversary. As this study argues, however, the surplus is not as much a question of numbers or quantities of arms or manpower as it is about the nature of existing military technologies and the likelihood of those technologies to change. Moreover, the surplus also depends on the reliable ability to detect cheating by an adversary, and the capacity to respond to that cheating through military means. On the definition, see Lipson, "International Cooperation in Economic and Security Affairs."

79. Michael Krepon and Dan Caldwell, eds., *The Politics of Arms Control Treaty Ratification* (New York: St. Martin's Press, 1991).

80. Feld, "Inspection Techniques of Arms Control," 320–21.

81. . I am grateful to John Owen and Jeff Legro for this point.

82. Michael Krepon and Mary Umberger, eds., *Verification and Compliance: A Problem Solving Approach* (Cambridge: Ballinger, 1988).

83. What I term "security environment" thus incorporates both offensive and defensive concerns. See Charles L. Glaser, "When Are Arms Races Dangerous? Rational Versus Suboptimal Arming." *International Security* 28, no. 4 (2004): 52–56.

84. The security environment plays an important role as a parameter in this study precisely because it is really the driving force behind both offensive and defensive realist explanations. For MET, the security environment plays a more important role in explaining the decisions to invest in the development and deployment of weapon systems, and less about the actual outcomes of subsequent cooperation that may seek to limit or reduce those weapon systems. Revisiting the prevailing security environment in each of the case studies that follows, allows us to control for political factors, and directs our focus to the influence of technology and technological change on states and their leaders.

85. George W. Downs and David M. Rocke, "Tacit Bargaining and Arms Control," *World Politics* 39, no. 3 (1987); Glaser, "Political Consequences of Military Strategy"; Glaser, "Realists as Optimists"; Robert Jervis, "Arms Control, Stability, and Causes of War," *Political Science Quarterly* 108, no. 2 (1993); Montgomery, "Breaking out of the Security Dilemma," 322.

86. Mearsheimer, "The False Promise of International Institutions."

87. Glaser, "Political Consequences of Military Strategy," 498–99; Charles L. Glaser, "The Security Dilemma Revisited."

88. Glaser, "When Are Arms Races Dangerous?," 45–47.

89. Security in this model is viewed as essentially an "absolute gain." Glaser, *Rational Theory of International Politics*, 132–35. For a debate on this issue, see Charles L. Glaser and John C. Matthews III, "Current Gains and Future Outcomes," *International Security* 21, no. 4 (1997). On relative gains, see Joseph M. Grieco, "Anarchy and the Limits of Cooperation: A Realist Critique of the Newest Liberal Institutionalism," *International Organization* 42, no. 3 (1988); Robert Powell, "Absolute and Relative Gains in International Relations Theory," *American Political Science Review* 85, no. 4 (1991).

90. Glaser, "Political Consequences of Military Strategy," 499; Glaser, *Rational Theory of International Politics*, 60–63; Glaser, "Realists as Optimists," 59, 71–72.

91. Peter Liberman, "The Spoils of Conquest," *International Security* 18, no. 2 (1993).

92. Charles L. Glaser, *Analyzing Strategic Nuclear Policy* (Princeton, NJ: Princeton University Press, 1990), 112, 19–24; George W. Rathjens, "The Dynamics of the Arms Race," in *Arms Control: Readings from Scientific American* (San Francisco: W. H. Freeman, 1973), 184.

93. Samuel P. Huntington, "Arms Races: Prerequisites and Results," in *The Use of Force*, ed. Robert J. Art and Robert Jervis (Lanham, MD: University Press of America, 1993); Rathjens, "The Dynamics of the Arms Race."

94. Glaser, *Analyzing Strategic Nuclear Policy*, 150–55; Glaser, "Realists as Optimists," 65–66; Van Evera, *Causes of War*.

95. Glaser, *Analyzing Strategic Nuclear Policy*, 150–55; Jack Levy, "Declining Power and the Preventive Motivation for War," *World Politics* 40, no. 1 (1987).

96. Glaser, *Analyzing Strategic Nuclear Policy*, 123–24; Glaser, "Realists as Optimists," 86.

97. Glaser, *Analyzing Strategic Nuclear Policy*, 44–49, 112; Jervis, "Arms Control, Stability, and Causes of War," 242–43.

98. Glaser, "Political Consequences of Military Strategy," 525–26; Glaser, "Realists as Optimists," 68–69.

99. Glaser, "Political Consequences of Military Strategy," 498–99; Jervis, "Arms Control, Stability, and Causes of War," 244.

100. Downs and Rocke, "Tacit Bargaining and Arms Control"; Downs, Rocke, and Siverson, "Arms Races and Cooperation"; Glaser, "Realists as Optimists," 56, 66–67; Jervis, "Arms Control, Stability, and Causes of War," 248–49.

101. Downs, Rocke, and Siverson, "Arms Races and Cooperation," 129–30, 38; Glaser, "Realists as Optimists," 65–66; Jervis, "Arms Control, Stability, and Causes of War," 240–42.

102. Glaser, "Realists as Optimists," 69–70.

103. Glaser, "Political Consequences of Military Strategy," 530; Glaser, "Realists as Optimists," 59–60; Jervis, "Arms Control, Stability, and Causes of War," 241.

104. See, for example, Robert Keohane, *After Hegemony*, 85–109.

105. See for example Richard Rosecrance and Arthur A. Stein (eds.), *The Domestic Bases of Grand Strategy* (Ithaca, NY: Cornell University Press, 1991). For an overview of liberal international relations theory, see Mark Zacher and Richard Matthew, "Liberal International Theo-

ry: Common Threads Divergent Strands," in *Controversies in International Relations Theory, Realism and the Neoliberal Challenge*, ed. Charles W. Kegley Jr. (St. Martin's Press: 1995), 107–50.

106. Richard Rosecrance and Arthur A. Stein, "Beyond Realism: The Study of Grand Strategy," Chapter 1 of Rosecrance and Stein (eds.) *The Domestic Bases of Grand Strategy*. Jeffery W. Legro, *Rethinking the World: Great Power Strategies and International Order* (Ithaca, NY: Cornell University Press, 2007).

107. Jack Snyder, *Myths of Empire: Domestic Politics and International Ambition* (Ithaca, NY: Cornell University Press, 1991).

108. Richard Rosecrance, *Rise of the Trading State: Commerce and Conquest in the Modern World* (New York: Basic Books, 1987).

109. This is the foundation of the liberal or democratic peace argument. See, for example, John M. Owen, *Liberal Peace, Liberal War: American Politics and International Security* (Ithaca, NY: Cornell University Press, 1997); Michael W. Doyle, "Kant, Liberal Legacies, and Foreign Affairs," *Philosophy and Public Affairs* 12:4 (Fall 1983); and Bruce Russett, *Grasping the Democratic Peace: Principles for a Post–Cold War World* (Princeton, NJ: Princeton University Press, 1993).

110. Matthew Evangelista, *Innovation and the Arms Race: How the United States and the Soviet Union Develop New Military Technologies* (Ithaca, NY: Cornell University Press, 1988).

111. See, for example, Bernard Brodie, "On the Objectives of Arms Control," *International Security* 1:1 (Summer 1976): 17–36; Frederich M. Scherer, *The Weapons Acquisition Process: Economic Incentives* (Cambridge: Division of Research, Harvard Business School, 1964).

112. This is the argument of Scott Allen, "A Comparison of the Washington Naval Arms Limitation Treaty of 1922 and the Strategic Arms Limitation Agreements of 1972" (PhD dissertation, University of Hawaii, 1996).

Chapter 2

The Washington Naval Conference

Much has been written about the Washington Naval Conference and its impact on the subsequent history of East Asia.[1] The three major Pacific powers, the United States, Japan, and Great Britain concluded a series of agreements, incorporating the concerns of other relevant powers, to fundamentally reshape the political environment in the region. The Washington Conference is generally viewed as a success precisely because it improved the security of the three powers by decreasing the probability of conflict through formal negotiation and the settlement of difficult issues that could have, if left unaddressed, precipitated a war in the Pacific.[2]

This chapter will focus on the Five Power Treaty of the Washington Conference and the expectations of existing and emerging naval technologies that shaped the bargaining positions of the delegates from the United States, Great Britain, and Japan and contributed to a successful arms control agreement.[3] The argument of this chapter, in following with the predictions of MET, is that the statesman and military leaders were able to exploit an opportunity provided by the prevailing nature of military technology to forge an agreement that could support political reconciliation and dampen potential conflict. The dreadnought battleship, which had been viewed as a revolutionary application from its inception two decades before the Washington Conference and was the index of naval power, had effectively reached a point of stable, if not diminishing returns.[4] While the big-gun battleship remained the "backbone" of any modern navy, the war had shown that it had achieved its maximum capabilities and it was unlikely to provide anything more than incremental improvements over time.[5] This perception of technological stability, and the relative certainty that battleships were not going to provide decisive new capabilities that could overturn a negotiated status quo, enabled statesmen to utilize them as bargaining assets to achieve a settlement. Ultimately, the negotiation of a 5:5:3 ratio (or 10:10:6 for the United States, Great Britain, and Japan, respectively) of battleships combined with the agreement of the United States and Great Britain to forego the development of specific fortifications in the western Pacific created a security environment in the region that the three major powers could confidently accept as an improvement to the existing situation.

At the same time, the parties at Washington were *not* able to cooperatively limit including aircraft, and their carriers, smaller surface vessels (most importantly cruisers), and submarines. The first two classes seemed to offer potentially decisive capabilities in the future. The third was seen by the smaller powers as a particularly useful hedge against the greater surface warfare capabilities of the larger powers. While their respective experiences during the Great War had provided military planners with some sense of their future impact on naval warfare, it was clear that their cooperative limitation would have been exceedingly difficult precisely because of their potential future value and because of the relatively inchoate stage of their development and difficulties of confidently monitoring their limitation. Examining the debates that were taking place within the respective naval establishments, as MET would predict, there was an emerging consensus that these assets offered potentially dramatic new capabilities and that further experimentation, research, and development were essential to national security over the longer term.[6]

Despite its shortcomings, the United States and Great Britain, the Washington Conference allowed for reconciliation between the erstwhile allies and dampened what was developing into an intense naval rivalry between the two powers, though the issue of cruiser would remain a challenge for the next decade.[7] Imperial Japan accepted constraints on its capacity for future expansion, but was rewarded with regional naval superiority and an effective political acknowledgment of Japanese preeminence in East Asia. With these benefits and the signing of the treaty also came a reversal of the dramatic and troubling deterioration of U.S.-Japan relations that had been underway at least since the end of the war.

The remainder of this chapter is divided into four sections. In the next section, I will briefly review the international political-military context leading up to the Washington Conference. In the second section, I will examine the respective views of the three powers concerning the dreadnought battleship and other key military technologies under consideration at the time, including submarines, aircraft, and other surface craft such as carriers and cruisers. In doing so, I will reconstruct the expectations that relevant leaders developed relative to the respective military and strategic value of these applications. The third section will revisit the Washington Conference and the negotiations that contributed to the successful arms control agreement codified in the Five Power Treaty as well the replacement of the Anglo-

Japanese alliance with the Four-Power Treaty. It will also briefly consider the aftermath and implications of the Conference. A brief conclusion will consider the history of the Washington Conference against the predictions of MET and alternative theories.

THE POST-GREAT WAR NAVAL SECURITY ENVIRONMENT

The political context in which the Washington Conference took place was characterized by high uncertainty and emerging rivalry. The Great War had eliminated the threat of German hegemony over the European continent, but at a high cost in blood and treasure for its participants. Nonetheless, the conclusion of the war and the harsh realization of its costs did not create a situation of peace free of troubles or potential rivalry. Two rising powers, the United States and Imperial Japan had emerged from the war in strong positions, but viewed each other with increasing suspicion, if not hostility. Great Britain viewed both of these powers as potential challengers to its traditional global naval preeminence. The growing U.S. Navy, and the commitment of subsequent U.S. administrations to increase its capabilities, threatened to usurp Britain's naval dominance. Japan, a formal ally of Great Britain, also created serious concerns, given its formidable naval capabilities and seeming intention to expand its empire in the Pacific, potentially threatening British dominions. The Anglo-Japanese alliance provided some insurance against this concern, but only complicated Britain's relations with the United States given the brewing security dilemma in the Pacific. Thus, the relations among the three remaining great naval powers were troubled and seemingly deteriorating.

The signing of the Treaty of Versailles may have formally brought the Great War to a conclusion, but it had left several important diplomatic questions left unanswered. Chief among them was the relationship between the United States and Great Britain.[8] While the U.S. entry into the war had certainly benefited the Allied cause and proved decisive in forcing the German surrender, troubles existed within the Anglo-American relationship. Under the strains of the war, and in attempting to bolster its blockade against Germany, the Royal Navy had imposed its naval superiority on the United States, even detaining U.S. merchant ships while America maintained official neutrality in the war. While obscured by the subsequent German actions of unlimited submarine war and the infamous Zimmerman Telegram, these British actions had a major impact on U.S. military and diplomatic thinking,[9] and directly led to the commitment by the Wilson administration to build a navy "second to none," as evidenced by the 1916 Naval Building Program and the subsequent 1918 Plan.[10]

Following the cessation of hostilities, Wilson, understanding the resource and fiscal constraints confronting Lloyd George's government, brandished the prospect of U.S. naval building to leverage support for the League of Nations.[11] Wilson also sought a commitment to freedom of the seas under the second of his famous Fourteen Points, a not-so-veiled threat to Great Britain's global maritime predominance and a direct assault on its naval policies during the war.[12] Moreover, Wilson's insistence upon British repayment of its large war debt only underscored the troubling state of British fiscal affairs. The "Battle of Paris" of March 1919 between Wilson and Lloyd George and their representatives over the definition of "freedom of the seas" and the future of British naval preeminence in the face of the American challenge was settled with an effective "truce" on building and a more realistic (and thus agreeable) interpretation of the second of Wilson's Fourteen Points.[13] This agreement would prove temporary, however. The subsequent refusal of the U.S. Senate to ratify the Versailles Treaty and the revolt against the League of Nations would dramatically undercut Wilson's plan for the postwar world and dramatically hamper diplomatic relations for the remainder of his time in office.[14] From Wilson's demand for "freedom of the seas" to incoming President Warren G. Harding's commitment to a U.S. navy "second to none," America seemed intent on usurping Britain's naval preeminence.[15] This exacerbated British perceptions of vulnerability derived from war exhaustion and dire resource constraints that would severely limit its ability to compete with any concerted U.S. challenge.

Britain emerged from the Great War with the largest navy in the world, but its disparate imperial commitments presented a number of problems that required attention. At the same time, England's domestic economic situation was grim. Demobilization and unemployment combined to put pressure on the government to allocate already constrained budgetary outlays toward social programs. During the war, defense expenditures could be justified, and the government, as would be expected, incurred debt and liquidated various holdings to maintain that war effort. The extremely difficult fiscal situation presented successive postwar governments with little budgetary flexibility and facing a public with little appetite for defense-related spending.

The primary strategic concern was the expansion of American capabilities, especially the naval building programs. Given the perceptions of American potential power, bolstered by the increasingly prevalent role of the United States as the primary creditor nation to its wartime allies, military and political leaders in London became increasingly focused on the Anglo-American relationship. The United States emerged as the one power capable of displacing Britain's global position. This unpleasant reality had already contributed to the replacement of Britain's traditional two-power standard with a more feasible goal of maintaining a one-power standard in naval power. This fundamental change in naval policy certainly reflects the acknowledgment of resource constraints, but it also underscores the commitment of Lloyd George's government and the broad consensus of Britain's political-military leadership to maintain its global naval capabilities.[16]

Facing the challenge of maintaining its vast overseas empire in the context of a war wary domestic political environment unlikely to support major spending on naval projects, British leaders were keenly interested in the prospect of a cooperative solution to the problem of competitive naval armament. Moreover, the timing seemed ripe for a reevaluation of relationships in the western Pacific that could secure British interests and also contribute to the improvement of Anglo-American relations, thus sustaining Britain's global position. However, when initial cooperative signals were seemingly rebuffed and the United States instead seemed committed to complete its expansion plans, the Cabinet approved the building of four new battle cruisers in March 1921. It is clear that a full-blown naval rivalry with the United States was something that policymakers in London feared above all. Losing an arms race to the United States would dramatically undermine British global power. Moreover, the likely impact of a prolonged naval race would be to fuel American hostility and increase the probability of conflict between the two powers. Thus the British preferences were to use the negotiations and improve, or at least maintain, positive relations with the United States while also maintaining a relative parity in capital ships, and thus the requisite capabilities deemed necessary to preserve its global position.

While Anglo-American relations were strained, relations between the United States and Japan were clearly deteriorating to the point where war had become a distinct possibility by 1920.[17] Under the first Wilson administration, U.S.-Japanese relations had become strained over the U.S. actions to maintain the "Open Door" to China for U.S. commercial interests, and the rise of anti-Japanese popular sentiment that led to the implementation of discriminatory legislation in California.[18] Japan's declaration of war on Germany and its rapid seizure of the Marshall, Caroline, and Marianas, only exacerbated the tense state of affairs and heightened U.S. fears of Japanese expansion.[19] This southward advance threatened America's lines of communication from Hawaii to the Philippines. Finally, the Japanese intervention in Siberia after the Bolshevik seizure of power precipitated the U.S. intervention at Archangel and Vladivostok with the latter operation specifically undertaken to monitor Japanese actions and decrease the likelihood of an opportunistic acquisition by the Imperial Army.[20] From almost 1906 forward, in the aftermath of Japan's decisive naval victory in the Russo-Japanese war, naval planners in the United States and Japan had been developing war plans and force structures that viewed each other as the most likely enemy in a future hypothetical conflict. In their earlier manifestations, these were relatively abstract formulations, but by 1914 these exercises had become more concrete and deliberate.[21] Public support for building a navy "second to none" combined with the seeming persistence of the American commitment to maintain the open door and its perceived interference in Japanese expansion created a diplomatic situation by 1920 that could be justifiably described as a "war scare."[22]

Unsurprisingly, Imperial Japan was highly concerned with U.S. plans to expand its naval capabilities. The standard of an 8-8 fleet (eight dreadnought battleships and eight large battle cruisers) was developed by the Imperial Japanese Navy (IJN) to be the basis for any realistic prospect of maintaining relative parity with the United States in the Pacific, though like its counterparts, fiscal limitations had made achieving such a benchmark exceedingly tenuous. Nonetheless, the Japanese government remained committed to the 8-8 fleet in the absence of any restraint on U.S. building and consistently interpreted American intentions in the region through the prism of Washington's publicly articulated naval building programs.[23]

The erosion of U.S.-Japanese relations was not lost on Britain. Though Japan was its formal ally, given the IJN's expanding naval capabilities and seeming intention to expand its influence in the Pacific, British dominions could be threatened.[24] The Anglo-Japanese alliance provided insurance against this concern, and allowed Britain to avoid assuming the costs of a maintaining a large Pacific fleet, but the alliance only further complicated Britain's relations with the United States given the brewing security dilemma in the Pacific.[25] The leaders of the Pacific Dominions had increasingly viewed Japan with suspicion and concern. Since the renewal of the alliance in 1911, it was viewed more as a tool of restraining Japanese expansion, particularly against British interests in the region. The respective positions of Australia and New Zealand in the southwestern Pacific made the Anglo-Japan alliance "mandatory" to the security of the Dominions and British holdings like Singapore and Hong Kong.[26]

The political environment during the period leading up to the Washington Conference was highly uncertain, with real potential for a naval conflict in Pacific. Despite exhibiting primarily security-seeking preferences, the United States, Great Britain, and Japan all seemed to have significant difficulties in interpreting the intentions of their counterparts and thus resorted to discerning motivation from more concrete actions like naval building. In the case of the Atlantic powers, it may indeed be true that neither of the Atlantic democracies desired war with each other. Great Britain may not have sought to forcibly maintain it naval primacy and the United States may not have sought to forcibly challenge Britain's position. However, neither was willing to simply accept a position of marked inferiority and the likely diminishing of political and diplomatic influence that would accompany such a situation. Despite the activities of those working for a peaceful solution on both sides and even with public pronouncements in support of good relations, naval capabilities—both deployed and planned—spoke far louder than any words.[27] The U.S. building programs, even with the existence of significant Congressional opposition, was a signal of U.S. commitment to increase its naval capabilities. Political and military leaders in London and Tokyo could only deduce from this behavior U.S. intentions to expand its influence.[28] As naval powers facing resource constraints, this could only be interpreted as a deeply troubling turn of events.

NAVAL EXPECTATIONS AND TECHNOLOGICAL CHANGE

The dreadnought revolution had fundamentally transformed naval warfare.[29] However, by the time of the Washington Conference, the consensus among naval and political leaders was that the maximum capabilities had been achieved and the dreadnought was no longer perceived as a revolutionary application. Through the war, it had become clear to policymakers that the all big-gun, heavily armored battleship and the formidable battle cruiser were no longer the cutting edge applications of naval warfare. They would indeed continue to be the "backbone" of a first-class naval power, and naval strength may continue to be measured in terms of capital ships for some time, but this perception was slowly changing. The dramatic advances in firepower, armor and speed initially incorporated in the dreadnought battleship (and the battle cruiser) were, by 1920, had diffused and were utilized in other classes of vessels such as cruisers, which proved more versatile and flexible for a wide array of naval missions (including those other than actual fleet combat). The prospects of naval aviation and the exploitation of airpower in a future naval war were also demanding greater attention. Perhaps most importantly, the Battle of Jutland, the single major fleet engagement of the Great War, underscored one fairly straightforward criticism of prevailing naval thought: the primary contribution of the battleship was limited to the engagement of other battleships.[30]

THE DREADNOUGHT BATTLESHIP: DIMINISHING RETURNS AFTER JUTLAND?

At Jutland, the Royal Navy's High Seas Fleet and the German North Sea Fleet fought a pitched battle on May 30–June 1, 1916.[31] The German fleet, including its core of dreadnought battleships, attempted to break out of its protected harbor in the Scapa Flow to the North Sea and ultimately the Atlantic to undertake commerce raiding operations against English merchant shipping. The High Seas Fleet was intent on keeping the German fleet bottled up, but because of the threat of German submarines, the battleships remained well-outside of effective range until the North Sea Fleet's gambit to escape from Scapa. Still, despite the hastened voyage of the German forces straight into the line of British ships, allowing the British to maximize their firepower upon their German targets, the result of the battle proved to be a stalemate. While the Germans were successful in sinking more British ships, most notably several British battle cruisers, ultimately the German fleet retreated back to the safety of its bases at Scapa and would not challenge the British again.[32]

In the wake of Jutland, Germany shifted its naval strategy, unleashing a campaign of unrestricted submarine warfare against British merchant shipping. This proved far more devastating to the British war effort than any pitched fleet battle, and while Jutland became a cause célèbre in Great Britain, the cumulative effect of the perceptions of battleship vulnerability and submarine effectiveness would spur debate on the future disposition of naval power.[33] Moreover, operations in the Dardanelles, where three British battleships were sunk by mines before the fleet could enter range to begin coastal bombardment had exhibited another major weakness in the battleship.[34] Even in the general media, the prevailing wisdom for continuing massive investments in dreadnoughts was under attack.[35]

Within the British government there was already a reassessment of the relative worth of the battleship going on.[36] Britain had stopped building capital ships during the war because of other pressing needs, and only had one battle cruiser, the *Hood*, under construction. Given the perception of the battleship as the primary naval warfighting application (and its limited utility for any other missions), American investment in these massive ships would improve their ability to prosecute a naval war in the near future. This increase in naval capabilities during peacetime intensified British concerns about American intentions.

> The Admiralty's case was based on the fact that if Britain built no new capital ships she would be 1925 still possess only one post-Jutland ship of that class (the Hood), while the United States fleet would have been strengthened by the six South Dakota class battleships and the six Lexington class battle-cruisers of the 1916 programme; while Japan would have completed four Kaga class battleships and four Akgai class battle-cruisers. Furthermore all the new American and Japanese ships were more powerful than the Hood, being armed with 16-inch instead of 15-inch guns.[37]

At the same time, British leaders felt increasingly vulnerable. Lloyd George and his closest advisors understood quite clearly that an arms race against the United States was unlikely to end in Britain's favor. Nonetheless, while Lloyd George preferred a cooperative solution, and had attempted to signal that fact by engaging in unilateral restraint after the war, he could not ignore the demands of the Admiralty to respond to the American challenge. For this reason, he signed off upon the plan to build four battle cruisers in 1921.

Within the Lloyd George Cabinet and the Admiralty, the challenge of maintaining Great Britain's naval preeminence was debated. While the goal of preserving Britain's position as a great naval power was not in question, the preferred capabilities and strategies to achieve that goal remained to be decided. The Admiralty represented by Admiral David Beatty and supported by Winston Churchill supported plans for new naval building, specifically an increase in capital ships. First, they remained hopeful of the possibility of some sort of cooperative agreement between the United States and Great Britain to avoid a full-scale naval arms race. Second, the civilian ministers were far less impressed with the perceived importance of the dreadnought battleship to naval security. This underlying tension between the civilian ministers and the naval leadership on the role of the battleship would remain a central point of contention throughout the deliberation leading up to the Washington Conference.

No single event conditioned the debate over the value of the capital ship among British political and military leaders more than the Battle of Jutland. The inconclusive outcome of the only major naval battle of the war seemed to imply that the combination of larger guns and superior armor had effectively outstripped tactical skill and maneuverability in major fleet engagements. Moreover, subsequent analyses and memoirs only underscored the perception of submarines as a threat to the fleet.[38] For example, the British Commander, Admiral Jellicoe, essentially admitted that concerns for German submarines influenced his decision to not pursue the retreating German Fleet. This admission only played into the narrative of leaders who were already questioning the battleship's value and its relevance moving forward.[39]

From a political (and budgetary) point of view, the more damning implication seemed to be that far less expensive applications like submarines and mines could severely conscribed the effectiveness of capital ships in wartime. Combined with the fairly inconsequential contributions of battleships in other theaters, it was difficult to maintain an uncritical commitment to further capital shipbuilding in the context of fiscal constraints and international uncertainty. A related consideration that mitigated the enthusiasm for the role of the capital ship was the undeniable contribution of other vessels during the war. In the Battle of the Atlantic, success was placed squarely on the role of destroyers in turning the tide of the unrestricted submarine warfare unleashed by Germany in the wake of Jutland. In fact, the decision of the United States to divert resources from their own capital ship program to the large-scale, rapid production of destroyers to protect convoys was a critical and timely one that resulted in a major shift in the Allies' prospects. Conversely, submarines had proven a formidable weapon that presented real challenges to the British war effort. While the influx of U.S. destroyers and convoy tactics ultimately defeated the submarine menace, it was not before the unrestrained utilization of submarines had taken a major toll on British shipping. Part of the necessity for the U.S. destroyer program was precipitated by the unwillingness of the Royal Navy to shift its destroyers to merchant protection in the Atlantic for fear that such a transition would make the capital ships in the North Sea vulnerable to torpedo attack. While battleship proponents in the Admiralty were often dismissive of the impact of the submarine, civilian leaders were much more cognizant of its role in the war.[40]

Similarly, in various Congressional hearings on naval appropriations and disarmament during 1920–1921 in the United States, there was significant disagreement concerning spending priorities.[41] In the fall of 1920, the General Board presented a three year building plan that would add three battleships and a battle cruiser among 88 vessels.[42] The focus on improving the U.S. inventory of battleships and battle cruisers in these building programs is not surprising. Given the quantitative disadvantage relative to Great Britain, and considering the perceived security requirements facing the United States, particularly the Japanese threat in the Pacific, the priority of achieving parity or superiority in these vital applications seems prudent. The capital ship remained the index of naval power, and no state with aspirations of maintaining a credible naval capability as a component of their national security strategy could do so without a fleet of these vessels. Nevertheless, it is also clear that within the military and political leadership of the United States, the prevalence of the capital ship over the longer term was in question. Several factors contribute to this development. First, the experience of the capital ship during the Great War was less than decisive. While U.S. battleships never engaged the enemy, the controversy that arose over the role of capital ships at the Battle of Jutland between the British and German fleets created serious questions about the naval importance of the capital ship. The British experience at Jutland implied that the actual utilization of the capital ship could be seriously limited by the presence of submarines. Moreover, when the two fleets actually engaged each other, despite pounding each other with their long-range guns, the battle ended in a stalemate. Adding the far more important role played by U.S. destroyers in reversing the effects of unlimited submarine warfare, and experts were left to question the real impact of capital ships to the winning of the war.[43] Considering the fact that the U.S. battleships remained at anchor during the conflict and it was less certain that battleships were the decisive weapon of naval warfare, particularly moving forward.

Second, while it may have been less of a concern in the United States than in fiscally constrained Great Britain, the per-unit costs of these large vessels was only increasing over time.[44] The technological leap that had been incorporated into the original dreadnought was no longer providing dramatic improvements in naval capabilities. The initial combination of unprecedented firepower, protection, propulsion, and range had made the earlier dreadnought vastly superior to preexisting battleships. By this period, however, this was no longer the case. Guns got bigger and armor was made thicker, but these were only marginal improvements, and the difference in capabilities between a dreadnought and a super (or post-Jutland) dreadnought was only one of degrees.

Finally, and most importantly, as U.S. political and military leaders looked forward, there were new applications that offered the potential to provide dramatic new capabilities and, in doing so, severely limit the utility of the capital ship. Most importantly, airpower seemed to present a truly revolutionary military application with the potential to transform warfare generally and naval warfare specifically. The use of aircraft during the war had only scratched the surface of its potential in warfare, and it was already viewed as an important auxiliary to fleet actions. While more optimistic proponents of naval aviation saw dramatic possibilities over the longer term, leaders were placed in the difficult position of requesting budgetary support for both preparedness and development.

Like its counterparts in the United States and Great Britain, the leadership of the IJN viewed the dreadnought battleship as the primary naval asset of the day. Japan had little direct involvement in the Great War aside from its operations against Germany in the Pacific, and support operation in the Mediterranean. Despite the effective stalemate that resulted at Jutland, the IJN remained firmly committed to improving its battleship capabilities.[45] In the 1917 building program, the Diet appropriated funds for two new "super dreadnoughts" or "post-Jutland" battleships, as they would be referred to at Washington. The *Nagato* (completed in 1920) and the *Mutsu* (completed 1921) were massive, displacing 32,720 tons, carrying eight 16-inch guns, and capable of achieving a speed of 26.5 knots. While their armor provided somewhat less protection than much slower American ships, the combination of speed and firepower placed these ships at the leading edge of capital ship capability.[46] The *Mutsu* would become a major issue at the Washington Conference, given that it was in the final stage of completion and initially targeted for removal under the initial American proposal. Two more battleships, the *Kaga* and *Tosa*, were scheduled for construction in the 1917 program but their funding was delayed until the passage of the 1920 building plan. These ships would have displaced over 38,000 tons and carried ten 16-inch guns and greater speed (despite their size) than their predecessors. In addition, the 1917 program included two battle cruisers the *Amagi* and *Akagi*, which when completed would displace 40,000 tons, carry ten 16-inch guns, and reach 30 knots. The construction of these massive battle cruisers was also delayed until December 1920.[47]

The focus of Japanese naval building during this period was driven by the perceived need to improve its capabilities in relation to the U.S. and British navies in the short term.[48] The latter possessed the largest navy in the world and the most formidable fleet of capital ships. The United States, in seeking to surpass British naval power, had focused on improving its capital ship inventory. Thus, in an increasingly competitive environment, to maintain pace with them, the IJN would similarly have to expand its own stock of battleships and battle cruisers. These ships were the index of naval power, and while other classes of surface vessels carried out important missions, prevailing wisdom dictated that battleships were necessary to winning a naval war through decisive fleet engagements. While Japan lagged behind the United States and Great Britain in terms of quantity of these ships, the IJN viewed the building plans of the period prior to the Washington Conference as means to not only close the existing gap in numbers of ships, but also to achieve a qualitative edge. Thus the *Nagato* and the *Mutsu* and the ships that followed were not merely facsimiles of existing U.S. and British ships, but improvements upon existing vessels that could potentially provide the IJN with an advantage in a future conflict.

Persistent support for big-gun battleship does not reflect an irrational commitment or parochial service interests across the upper levels of the British, American, and Japanese naval leaderships.[49] Rather, in assessing *preparedness* needs versus investment in experimentation and development, there were logical and proper concerns voiced on both sides. What is clear is that the battleship was a known commodity by this time, and depending on the relative position of the state in comparison to the other naval powers, we would expect that leaders would seek to acquire these weapons to close any existing quantitative deficits. But looking forward, other platforms and applications seemed to be demanding further investigation if a navy was to maintain at the head of the technological curve.

NAVAL AVIATION: THE NEXT REVOLUTIONARY APPLICATION?

By 1920, the development of naval aviation and the perceived significance of carrier operations in future naval warfare dominated the discussions of military and civilian policymakers. This reflected a larger dialogue that was underway within the various branches of the military concerning the optimal methods to exploit what seemed to be an emerging revolution in technology.[50] Experience of the war had only seemed to "scratch the surface" of what airpower could offer. More than just scouting, and fire control missions, larger numbers of more robust craft could significantly impact future fleet engagements, and carriers may become the "capital ship" of the future.

Airpower advocates were enthusiastic about the role of airpower and viewed the impact of aviation in a variety of roles and missions as decisive, potentially changing the nature of future warfare. They also argued that airpower would be influential in the shorter term and pushed for funding for experimentation, development, and even deployment sooner rather than later. General Billy Mitchell, who had recently engaged in a series of experiments utilizing bombing planes against ships, was only the most vociferous and forceful of the advocates.[51]

Control of the air was viewed as the decisive factor in future naval engagements. Air battles would be decided by the force that was able to take the initiative, and bring overwhelming force to bear on the enemy.[52] Aircraft were clearly offensive weapons that would seek out and destroy the enemy's air force and the advantage would be with the attackers.[53] Once control of the air was achieved, the enemy fleet would be left vulnerable to attack. Thus, even if naval aviation was still viewed as an auxiliary to the big-gun battleships, it was a critical one, and could significantly influence the outcome of naval warfare. Over time, perhaps, the carriers would be equipped to be able to sink surface craft, but in the early stage of anticipating the types of engagements that would take place, speed was clearly the central attribute for carriers.[54] The timing for investment if experimentation and research was particularly beneficial in that the United States could leverage its own aviation development upon the significant progress that had already been made by the British. In fact, British experimentation had already succeeded in illustrating some of doctrinal and tactical issues that would be faced by U.S. aviators.[55]

Britain was at the forefront of this investigation, and in fact was already engaged in bureaucratic battles over the organizational structure of military airpower moving forward. The Admiralty was fighting to maintain autonomy of naval aviation against calls for centralization of airpower under a Royal Air Force. Nonetheless, the British had developed several variations of an aircraft carrier, tailored for a variety of missions, and had utilized airplanes for bombing missions against German naval assets in the lead-up to Jutland. While aircraft and their carriers may have continued during this period to be viewed as auxiliaries to the fleet (to which battleships and heavy cruisers were still central) they were increasingly viewed as an important component. Scouting, targeting of enemy ships, and bombing were important missions that aircraft attached to the fleet would play in future operations. Sir Hugh Trenchard, for example, argued: "If we can concentrate on the development of the aircraft, the importance of the capital ship must be greatly reduced and will become an insurance for which we cannot afford to pay the premiums."[56] Support for aviation and opposition to further capital shipbuilding came from both within and outside the Royal Navy. As Gamble explains:

> Stronger claims for air power were made by retired Major-General Sir Frederick Sykes, who stated: "All forms of warship will soon become obsolete, and by the time a programme of battleship construction is completed new factors may have arisen which render them useless." Retired Rear-Admiral Charles de

> Bartolome, wartime Third Sea Lord, strongly supported Sykes's views concerning the enormous potential of airpower. He asserted that new technologies "are so promising that we should be well advised to defer the building of capital ship."[57]

Despite the Admiralty's continued support for capital ships, the Royal Navy was also investigating the role of airpower and moving forward to explore its capabilities. Having developed two aircraft carriers during the war, the Royal Navy was already at the forefront of naval aviation development and with the added pressure of competition from the fledgling Air Ministry, experimentation progressed.

> In March 1921 Admiral Chatfield circulated a comprehensive paper by the Staff on the state of naval aviation in the Royal navy. The Staff surveyed the recent experience gained in the operation of aircraft from the Argus and the Eagle, and described the progress made in the development of specialized equipment such as catapults and arresting gear. They also listed the new types of aircraft ordered by the Air Ministry or projected for naval service, and the new weapons being designed for their use. This document provides convincing evidence that the Admiralty, though they regarded the big gun as the dominant and decisive weapon, was certainly not laggardly in supporting those aspects of naval aviation which lay within their control.[58]

These elements within the Admiralty strongly supported a greater commitment to innovation and experimentation on naval aircraft. Admiral Richmond argued before the Committee that "If the battleship programme hinders our research work it will do harm . . . merely to build ships of a larger and more heavily armed type than those of the United States does not solve the problem."[59] Looking forward, airpower and the ability to integrate naval aviation within fleet operations would be increasingly important.

The U.S. Navy had relatively little experience with aviation during the war, but technological trends and the desire to maintain pace with other states demanded that both aircraft and vessels that could carry them would become a priority. The United States found itself behind both England and Japan in this important area, and given the potentially decisive impact of airpower on naval operations, it was crucial to initiate programs to allow the U.S. Navy to catch up.[60] Throughout the Congressional hearings of 1921, questions regarding the development of naval aviation and the perceived significance of carrier operations in future naval warfare repeatedly dominated the discussions of military and civilian policymakers. This reflected a larger dialogue that was underway within the various branches of the military concerning the optimal methods to exploit what seemed to be an emerging revolution in technology.[61]

Obviously, if aircraft were important, then the carriers that brought them into battle would also a critical role to play and would have to possess certain necessary requirements. "We want the carriers to form floating airdromes that can act whether the sea is rough or smooth, that have a great radius of action and that can carry a tactical unit of planes."[62] Most importantly, they would need speed to keep up with the fleet and also be maneuverable and fast enough to avoid enemy battleships and other attacking surface vessels. Over time, perhaps, the carriers would be equipped to be able to sink surface craft, but in the early stage of anticipating the types of engagements that would take place, speed was clearly the central attribute for carriers.[63] The timing for investment if experimentation and research was particularly beneficial in that the United States could leverage its own aviation development upon the significant progress that had already been made by the British. In fact, British experimentation had already succeeded in illustrating some of doctrinal and tactical issues that would be faced by U.S. aviators.[64]

Mitchell was certainly not alone, however. Within the U.S. Navy, there were also advocates of airpower, concerned that not enough was being done to exploit this new military application. Rear Admiral Sims, the president of the Naval War College, also saw the critical importance of airpower and aircraft carriers. Having devoted significant time to working through many of the conceptual issues of utilizing aircraft carriers and naval aircraft in fleet engagements in war games at Newport, Sims saw great value in the carrier, particularly with the development of more reliable aircraft. Over time, the carrier would likely displace the battleship as the "capital ship" and the centerpiece of fleet operations. With a certain level of speed, and carrying aircraft capable of dropping bombs or torpedoes on enemy ships, the carrier would be able to effectively attack other fleets without ever coming within range of the guns of enemy vessels.[65]

If the U.S. Navy and the Congress placed priority on the development of carriers and aircraft and the various technological hurdles could be overcome, then the United States would be in a much better position. However, at the current time, with little progress made in this field, the U.S. Navy would become increasingly vulnerable, particularly in the face of states that were investing in and developing these capabilities. All things being equal, in a future war, a fleet with greater aircraft capabilities would face dramatic advantages over a fleet without such capabilities.[66] To underscore his arguments for increasing the U.S. naval aviation program, Sims presented a hypothetical engagement between two fleets. When comparing them, he made a straightforward mathematical analysis of the likely air battle to illustrate the importance of aviation in sufficient numbers.

> Well if some other fleet should have as much as, say, three times the number of airplane carriers that the British fleet has, then, of course the first effort task of this fleet would be to wipe out all the British airplanes out of the air. That would be the first thing necessarily. Now, if we assume that the opposing fleet

> loses as many planes as the other one does, then this fleet would have two-thirds of her planes left with which to bomb and torpedo the other fleet. If the assumptions that we have made are true, I think they would destroy that fleet. I regard it as a fundamental weapon.[67]

Contrary to arguments that countermeasures for airplanes would also inevitably be in the offing, Sims argued that once technical hurdles for planes were overcome and carriers assumed the role capital ships in terms of fleet planning and operations, air power would remain a central component of naval warfare for the foreseeable future. Naval aviation may develop functional specifications, such as separate bombing/torpedo and pursuit/fighter craft, but the technological premises upon which naval aviation are built would seem to imply that the development of credible countermeasures from surface vessels would be highly difficult. Sims argued with confidence that the technology of surface ships (specifically battleships) was unlikely to provide capabilities that could seriously undermine the perceived impact of naval aviation, specifically its ability to attack and sink surface vessels if control of the air has been established. "Now, if you have a greater air force than the other navy has, you can wipe all his planes out of the air, including his bombing planes, and then you would be free to do your bombing and torpedo firing afterwards."[68] At the same time, while Sims argued for the necessity of experimentation and development of aircraft and carriers, he (like Mitchell) was also sympathetic to the arguments of the General Board and the CNO concerning the need for battleships and cruisers in the current requests for appropriations. In the short term, it was only prudent to invest in assets that would be required to effectively carry out critical missions in a possible war. However, over time, it was important to revisit the issue and reevaluate the respective utility of the battleship as the capabilities of the aircraft carrier improve.[69]

Japan's navy had consistently proven to be concerned with exploiting technological advances to improve capabilities of the fleet. Aviation was one important area where the IJN was at the leading edge of experimentation and innovation. Japan had utilized seaplanes launched from a converted transport ship, in the operation to take the German base at Tsingtao in September 1914.[70]

> Viewing the air operation at Tsingtao as a whole, Charles Burdick, the acknowledged authority on the siege, has asserted that "the sophistication of the Japanese aircraft employment—i.e., coordination with land forces, bombing equipment, and general mobility—was well ahead of any other country."[71]

As early as 1915, elements within the IJN were agitating for greater focus on naval aviation and that "even though the airplane was in its infancy, it was destined to be the decisive weapon of the future, and the dreadnought battleship was now fatally threatened by the aerially launched torpedo and the mine."[72] However, while the Japanese Navy may have fostered the development of aviation earlier than the United States, its leadership followed the trend of its U.S. and British counterparts by prioritizing the construction of battleships in the short term to achieve the 8-8 fleet. Navy Minister Kato Tomosaburo echoed the views of leading U.S. and British naval leaders when, in response to questioning from the Diet regarding the impact of airpower on naval operations:

> he declared with confidence that there was as yet no aerial bomb that could put a warship out of commission. That didn't mean that the navy shouldn't make sure that in the future its warships were safe from aerial attack (by either bombs or torpedos), but that for the present, most of these dangers were largely theoretical. For that reason, Kato assured his listeners, the Japanese navy did not regard the advent of aircraft as upsetting the traditional dominance of the battle fleet.[73]

Kato's response reflects the pressing concern among the naval leadership to achieve the 8-8 fleet, and thus the four super dreadnought battleships deemed essential to maintaining Japan's security in response to the U.S. building program. His goal was finally realized in July 1920 when the Diet approved a massive building program.[74] While the centerpiece of the program was the four battleships, it also secured funding for the construction of submarines and an aircraft carrier. In addition, the program also included funds for the invitation of French and British pilots to serve as flight instructors for Japanese naval pilots.[75] These latter items underscore the increasingly important role of aviation within the IJN during the period leading to the Washington Conference, following a similar pattern exhibited in the American and British experiences. While the limited capabilities presented by existing aircraft precluded higher levels of investment, particularly in light of the perceived need to rapidly increase the surface capability of the fleet to address U.S. naval expansion, experimentation and organizational adaptation to develop aviation continued.[76]

Despite active debates with the leadership and planning circles of the U.S. Navy, these views reflect differences of emphasis. The questions were about placing priority on experimentation with aviation and carrier development; the leadership of the Navy was not opposed to carrier development. To the contrary, the development of carriers was viewed as a priority. While perhaps less enthusiastic about the short-term capabilities of carrier aviation and the displacement of the battleship by carriers, it is clear that aircraft and fast carriers were seen as an important auxiliary to fleet operations. The battleship remained the central application for fleet operations, but aircraft had already proven to be important in carrying out reconnaissance, fire control, and even bombing missions.[77] Preparedness dictated priority be place on "rounding out" the fleet and expanding the inventory of capital ships. Development concerns pushed for greater investment in carriers and aviation. The leadership of the U.S. Navy attempted to strike a reasonable balance between these two positions in the context of resource constraints.[78]

This discussion reflects the relatively widespread consensus among military and civilian leaders in the United States, Great Britain and Japan as to the utility and future role of the dreadnought battleship in naval warfare. For the short term, states could not presume to compete with other naval powers without some level of dreadnought battleships and battle cruisers in their inventory. However, over time, the relatively unchanging capabilities of these massive vessels would significantly circumscribe their contribution to the fighting and winning of future wars. Naval airpower, while still in its infancy, seemed to possess the potential to transform naval warfare. The development of more reliable aircraft and the vessels necessary to transport them into battle were still daunting challenges, but by the early 1920s, leading military and civilian leaders viewed them as attainable and a necessary focus of investment. The successful exploitation of aviation would provide dramatic new capabilities that could prove to dominate the nature of future war and, in doing so, severely degrade the utility of the battleship, if not drive it from the sea.

THE CRUISER: ACHIEVING THE TAIL-END OF THE DREADNOUGHT REVOLUTION

The firepower and armor improvements that had initially marked the dreadnought "revolution" were now being transplanted to smaller vessels, and thus there were significant concerns that continuing to spend scarce resources on larger and larger capital ships with limited utility beyond major fleet engagements was short-sighted. Smaller ships, with greater speed but formidable gunnery and armor, may prove to be far more useful for a variety of missions and still provide adequate firepower for fleet operations.[79]

> The different classes of Cruisers had evolved in response to several distinct functional needs. The original cruiser functions, performed in earlier times mainly by frigates and sloops-of-war, were commerce raiding and commerce protection. The latter function sometime involved the convoy of merchantmen through zones of special danger, sometimes the pursuit of enemy raiders. All of these operations required ships sufficiently large to keep the sea in any kind of weather, tolerably comfortable accommodations for officers and crew, fuel capacity for long voyages, sufficient speed to overtake merchant vessels and to escape hostile battle cruisers, some armor protection against gunfire, and batteries that could outshoot the enemy's commerce raiding cruisers.[80]

While carriers and naval aviation would ultimately prove to be the most decisive of the weapons considered for limitation at Washington, it was ultimately the cruiser question which left unresolved would create diplomatic headaches and prove difficult to resolve in the aftermath of the conference. The perceived naval benefits of the dreadnought revolution now increasingly favored smaller ships. They did agree to a proposal by Hughes to limit cruisers to a displacement of 10,000 tons and a maximum of 8-inch guns. "The conference adopted the proposal with little debate, for they did not want nations building auxiliaries that were capital ships in everything but name."[81] However, beyond this definitional limitation, and given the eventual agreed limitations on capital ships, cruisers only became more valuable in the eyes of British leaders.

> so wide is the field and so little known are the dangers which may threaten our merchant ships in 10 years' time that we cannot be bound by any restriction in regard to light craft of whatever suitable type we may require to build for this purpose.[82]

The Admiralty argued that they Royal Navy could use large numbers of relatively smaller cruisers for fleet operations and also to protect Great Britain's flow of commerce from the Commonwealth states and the Dominions. Without any limitations on submarines, auxiliary ships including cruisers and destroyers took on greater importance, and at the urging of Great Britain (concerned about French plans to build submarines) no limitations were implemented.

> In this respect Great Britain's position is unique, the U.S. broadly speaking does not rely on the sea at all for the feeding of her population and Japan to a far less extent than Great Britain and with comparatively short sea communications.[83]

Given its large network of naval bases around the globe, smaller cruisers would serve their purpose in a relatively cost-effective manner. The United States similarly saw an increased value in cruisers as a result of the limitations on capital ships but perceived a need for heavier boats, capable of greater range to make up for the lack of basing options. For its part, more hawkish elements within the IJN and the Japanese government agitated against any further limitations that would widen Japan's position of relative inferiority vis-à-vis the Atlantic powers. However, the real controversy emerged between Washington and London during the mid-1920s, as Great Britain laid down the keels of several new cruisers as the United States had forgone building in the context of tight budgets.[84]

SUBMARINES: DIFFERING VIEWS AND PERCEPTIONS

The development of the submarine, which had already proven to play a decisive, if controversial, role in the German war against allied shipping during the war, was deemed important by naval planners.[85] More sophisticated submarines would likely play a wide variety of roles in a future war, and as weaponry, maneuverability, and communication improved, these vessels, too, may prove to be devastating to the traditional capital ship. While Britain's experience understandably contributed to a consensus within the Admiralty and their political allies that the weapon should be prohibited, this was viewed as indirectly increasing the perceived value of their extensive fleet of surface vessels. In fact, while the Fisher "Revolution" is associated with the *HMS Dreadnought* and the battle cruiser *HMS Invincible*, he also saw subsurface warfare as a potentially decisive contributor to future naval warfare.[86] Similarly, in the spirit of "rounding out the fleet," the leadership of the U.S. Navy saw the future of the submarine as particularly useful and while they viewed American submarine development until that period as lagging beyond that of the Europeans, the General Board viewed larger, longer-range boats as a key research and development target.[87]

> The submarine as a man-of-war has a very vital part to play. It has come to stay. It may strike without warning against combatant vessles, as surface ships may do also, but it must be required to observe the prescribed rules of surface craft when opposing merchantmen as at other times. . . . As a scout it has great possibilities.[88]

Likewise, the IJN which had experimented with submarines and perceived them as a versatile and potentially decisive weapon opposed limitations. Perhaps the most proactive and forceful advocates of submarines were the smaller powers, France and Italy. Having accepted parity with each other in capital ships, they had also formalized a numerical inferiority to Great Britain, the United States, and Japan. Given this significant imbalance, neither power would consider limits on submarines which came to be viewed as relatively cost-effective alternatives to capital ships and formidable insurance in the event of a future naval conflict.[89]

These various currents in technological change shaped the expectations that military and civilian leaders developed concerning the impact of these various weapon systems in a future war, and clearly we can see the importance of time frames at work in relative importance of preparedness versus development. With this understanding of the military expectations of technological change held by naval leaders prior to the Washington Conference, we can now begin to understand how the successful limitation of capital ships was negotiated in the Five-Power Treaty and served as the basis for a major arms control success.

THE WASHINGTON CONFERENCE

The Washington Naval Conference opened with much fanfare and public and media interest on December 1921.[90] President Harding used the dedication of the Tomb of the Unknown to set the stage for discussions and to serve as a poignant reminder of the importance of the work that was being undertaken.[91] After brief remarks to open the conference, the chairman, Secretary of State Hughes took the podium and gave his speech. While most of the audience, including the respective delegations, had expected the Secretary's speech to follow the president's with further celebratory and inspirational language, Hughes shocked his listeners by proceeding to offer a highly detailed, technically informed plan for naval arms limitation.[92]

> In making the present proposal, the United States is most solicitous to deal with the question upon an entirely reasonable and practicable basis, to the end that the just interests of all shall be adequately guarded and the national security shall be maintained. Four general principles have been applied:
>
> (1) That all capital ship building programs, either actual or projected, should be abandoned;
> (2) That further reduction should be made through the scrapping of certain of the older ships;
> (3) That, in general, regard should be had to the existing naval strength of the Powers concerned;
> (4) That the capital ship tonnage should be used as the measurement of strength for navies and a proportionate allowance of auxiliary combat craft prescribed.[93]

Based on work that he had done primarily with the assistance of Assistant Secretary of the Navy Theodore Roosevelt Jr. and going far beyond the recommendations of the U.S. Navy General Board, Hughes presented a program of cuts that would dramatically reverse the trend in competitive building. Existing building programs, including the U.S. 1916 Program, would be effectively halted.[94] Vessels under development would be scrapped. Further cuts would be made in existing inventories: older ships would be scrapped and not replaced. The plan was based on a premise worked out by Hughes and Roosevelt to first "stop now" and end existing building and then cut back on existing programs by removing vessels from service.[95] The respective naval powers would agree to maintain ratios, based upon tonnage displaced, and the combination of cuts and termination of building already underway would leave the United States and Great Britain with a rough parity, while

Japan would be relegated to approximately 60 percent of the U.S. and British levels, thus the plan was typically described as either 5-5-3 or 10-10-6 for the three powers. It was an ambitious plan, and precisely because of its scope, neither Hughes nor Roosevelt was certain that President Harding would accept it prior to the conference, but surprised them by enthusiastically endorsing it. The plan was received with great public and media support.[96] "One commentator remarked, 'Hughes sank in thirty-five minutes more ships than all the admirals of the world have sunk in a cycle of centuries.'"[97] Under the Hughes plan, "there would be immediately destroyed, of the navies of the three powers, 66 capital fighting ships, built or building, with a total tonnage of 1,878,043."[98]

Hughes's speech was viewed as a diplomatic coup. This "bombshell" not only clearly signaled the willingness of the United States to dramatically limit its own naval capabilities in order to forge a successful cooperative agreement; the speech also put the other delegations on notice. Given its inherent advantages, the United States had taken the initiative, and, in less heroic terms, had significantly decreased the bargaining room of its partners. Foot-dragging and obfuscation by other delegations would necessarily be viewed in the light of this dramatic and magnanimous proposal by the United States and thus the public relations battle was already decidedly in the favor of the U.S. delegation.[99]

The Committee on the Limitation of Armaments met twenty-one times between November 15, 1921, and February 3, 1922.[100] In fact, the agreement among the United States, Great Britain, and Japan was accomplished by December 16, 1921, when Secretary Hughes wrote to French Minister Briand, who had left the conference after its initial plenary session, to bring France and Italy into the agreement at a ratio of 1.75 each.[101] The French initially rebuffed this plan, making noises about the needs to maintain a large naval capability. Italy, conversely, was willing to accept parity with France. Despite early signals that France desired to achieve a navy that it had sacrificed for the war effort direct appeals to Foreign Minister Briand, acknowledging the inherent challenges confronting France regarding a major naval buildup, agreed to the proposed ratios.[102] Thus, in little more than a month, the challenge of limitation of capital ships had been resolved. However, this should not imply that the negotiations were not difficult.

FINDING AGREEMENT: LIMITS ON CAPITAL SHIPS AND ISLAND BASES

From the outset, while both the British and Japanese delegations had agreed to the plan "in principle," it was clear that the Japanese were concerned about the numbers. Baron Kato Tomosaburo bargained hard in favor of the 70 percent ratio. The Japanese Navy had established that as a necessary requirement for the defense of the empire against the United States. Interestingly, Secretary Hughes had a fairly clear idea of how far the Japanese would go. Throughout the conference proceedings, the secretary maintained close contact with Ambassador Warren in Tokyo, who in turn, remained in contact with the Japanese Foreign Ministry. In addition, the United States had successfully cracked the Japanese diplomatic code and were able to monitory transmissions between the delegation in Washington and the Foreign Ministry in Tokyo. This intelligence and the Hughes-Warren backchannel would provide significant leverage for Hughes. On the November 19, Hughes wrote with frustration to Warren that the Japanese commitment to a 70 percent ratio was effectively a rejection of cooperative arms control and would not represent any real reduction. As naval forces were constituted, Japan would actually have to expand to reach the 10-6 ratio, never mind a 70 percent relation to U.S. forces.[103] Warren persistently bolstered Hughes's negotiating position with reports that Tokyo was less committed to the 70 percent ratio than its representative.[104]

The Japanese delegation was reluctant to move off the commitment to 70 percent for the Imperial Navy and may have been a tactic to achieve U.S. concessions for a more pressing strategic problem for Japan the fortification of U.S. possessions in the western Pacific. Guam and the Philippines represented a clear threat to the Japanese security perimeter, and with the development of air power, the Japanese home islands could potentially be within range of U.S. forces in the event of a future conflict. This forward base capability had played a role in the Japanese decision to annex the German islands of Micronesia. Now, in return for Japanese agreement to lower their naval requirements, the United States may have to agree to forego the fortification of these islands to improve Japan's security at the expense of future U.S. power projection capabilities. Japan had signaled this tradeoff earlier, but Admiral Kato first broached the proposition on December 2. After making clear that "when he came to Washington he had arrived with a double determination, first, to obtain a real limitation of naval armaments, and, second, to secure a definite and permanent improvement in the relations of the two countries."[105] Nevertheless, he continued to negotiate a hard line on the ratio issue reflecting the centrality of the 70 percent ratio to Japanese Naval requirements, even if that objective had not been achieved prior to the conference. In addition, the post-Jutland dreadnought, *Mutsu*, had been completed prior to the conference and was mistakenly listed as an incomplete vessel and thus one to be scrapped. Accepting the lower ratio may be possible, but destroying a completed modern dreadnought, which the IJN had fought so hard to get funded, was extremely difficult to countenance.[106]

The 70 percent ratio was thus presented to be a minimum force level for capital ships to adequately provide for Japanese security. However, Kato shrewdly planted the seeds for a compromise that could break this impasse. He had noticed that whenever news was received in Japan about the possibility of fortifications on the American islands in the Pacific it had caused a significant public alarm. An agreement that would prohibit such fortifications or naval bases and maintain the status quo would serve Japan's interests and remove the need for a larger ratio of capital ships.[107] In fact, as Hughes had admitted to Balfour previously, the likelihood of significant development of U.S. military assets on Guam or in the Philippines was unlikely. The difficulty of obtaining adequate funding for the various naval appropriations bills since the 1916 program did not bode well for the fortification of bases. U.S. Navy planners had advocated on behalf of such secure forward bases if the United States was indeed committed to maintaining an "Open Door" to China, by force if necessary. However, it was unlikely that these would actually be constructed and Hughes privately doubted the underlying premise that the United States would actually go to war to prohibit or reverse Japanese expansion in China, even if such behavior limited American access.[108] During this time it became clear that Britain had adopted the view that a successful conference would represent an improvement to its security, and its representatives worked with Hughes to pressure the Japanese. During discussions on December 2, Balfour expressed his view that the Hughes proposal has requested sacrifices of all of the powers, but would improve their respective security situations.[109] Balfour's intervention on behalf of the Hughes plan increased the pressure on Baron Kato, and he essentially stalled by requesting time to consult with the foreign ministry in Tokyo. On December 3, Warren reported to Hughes that the Japanese may be willing to adjust the ratio if the fortifications issue could be successfully addressed.[110]

Here the conference nearly faced a major setback. Hughes responded with some frustration to this note from Warren, precisely because Admiral Kato remained committed to the 70 percent ratio. The Imperial Navy had not achieved that force level and would have to undertake significant building to achieve it. Such a view did not seem to correspond to a serious commitment to arms limitation. And thus if it was a race that Japan wanted, the United States could certainly accommodate them.[111] At the same time, however, Hughes did agree that the issue of fortification was one that could be cooperatively addressed and if such an agreement would facilitate the Japanese acceptance of a 60 percent ratio, the United States could support such a measure.[112] Nevertheless, Baron Kato maintained his hard line on the ratio and repeatedly requested time to communicate with his government, thus implying that it was not his personal initiative to obstruct an agreement, but reflective of large Imperial diplomatic policy. Warren reinforced Hughes's bargaining position with information about the Japanese government's view of the proceedings.[113]

Ultimately, Japanese officials saw the island fortification issue as more important that the maintenance of the higher ratio, which was more likely the preference of the IJN. The willingness of the United States to forego the development of forward bases in the Philippines and on Guam would significantly improve Japanese security by restraining the American ability to project power in the western Pacific and remove a real threat to the Japanese home islands. Thus the trade-off, limitations on capital ships—those deemed most important to actually fighting a naval war—and forward bases, would transform the security environment in the region and improve Japan's security. At the same time, Japan's willingness to limit it, naval forces would improve the security of the U.S. and British interests in the region. In addition the replacement of the Anglo-Japanese Treaty with a Four-Power Treaty (United States, Great Britain, Japan, and France) could remove the troubling diplomatic aspects of the alliance with a more consultative arrangement that would support peaceful resolution of c0onflicts and disputes that may arise between the powers.[114]

The stage for a compromise was set when on December 12, Baron Kato approached Hughes and Balfour and accepted the 60 percent ratio, but on the condition that the fortifications issue would be resolved in the context of the Five-Power Treaty. In addition, Kato requested that the Japanese be allowed to keep the post-Jutland battleship, the *Mutsu*. The IJN would scrap an older ship, the *Settsu* in its place. Both Hughes and Balfour accepted this change to the formula, but allowing Japan to keep the *Mutsu* did seem to underscore Britain's existing naval-building problem. Having ceased its building programs since the war, Britain would be left without any post-Jutland ships, except for the battle-cruiser *Hood*. For the next few days, the British considered different alternative reduction formulas that would allow for the building of new ships to offset the Japanese retention of the *Mutsu*, while maintaining the underlying ratio system. Ultimately, Balfour and Hughes were able to come to an agreement that would allow the British to build two new capital ships, in exchange for scrapping four older King George V model battle cruisers. Ultimately, Britain would have a slightly higher ratio level than had been previously ascribed to it in the original Hughes proposal, but both the United States and Japan accepted this. On December 15, the three delegations were able to announce an agreement to the assembled press:

> An agreement is reached between the three Powers—The United States of America, the British Empire, and Japan, on the subject of naval ratio. The proposal of the American Government that the ratio should be 5:5:3 is accepted. It is agreed that with respect to fortifications and naval bases in the Pacific region, including Honkong , the status quo shall be maintained, that is, that there shall be no increase in these fortifications and naval bases except that this restriction shall not apply to the Hawaiian Islands, Australia, New Zealand and the islands composing Japan proper or, of course, to the coasts of

> the United States and Canada, as to which the respective powers retain their entire freedom.[115]

This agreement would prove both a major accomplishment and also provide its own difficulties. The momentum toward a larger resolution was certainly welcome news to observers, but for the French delegation, exclusion from the "'big three" was a blow to French prestige. Hughes and Roosevelt's plan would relegate France and Italy to ratios of 1.75 respectively. As discussed above, both nations accepted the ratios on capital ships. While real limitations on capital ship tonnage and the agreement to forego fortifications of the western Pacific insular possessions had been achieved, the objective of extending limitations to other classes of vessels would prove difficult and ultimately impossible.[116]

NEGOTIATIONS TO LIMIT OTHER CLASSES OF WEAPONS

Aircraft carriers were limited, but in a way that would allow the states to continue to develop and experiment in these important applications. "No airplane carrier shall be laid down during the term of this agreement whose tonnage displacement is in excess of 27,000 tons, and no gun shall be carried by any such ship other than a capital ship in excess of 8 inches."[117] The ratios set for these were initially set to follow the capital ship ratio, but ultimately, an agreement was reached in which the United States and Great Britain would be allowed 135,000 tons, Japan 81,000, and France and Italy 60,000 tons each. Given their experimental nature, it was agreed that they could be replaced without reference to age.[118] More generally, the delegations were unable to cooperatively create limitations on aircraft because of the difficulty of the underlying problems such as the civilian development of aircraft.[119] Given the wide-ranging possibilities associated with progress in the development of aviation for mankind, the attempt to develop any sort of effective limitation on military aviation was difficult to comprehend. Ultimately, the only way to limit military aviation would be to limit all aviation, which was viewed as impossible and unrealistic."[120] Thus aviation was left to develop without restraints. This is not surprising given the inherent difficulties of the task and the perceived decisive nature of military aviation for the security of states.

Given the agreed limitations on capital ships, the extension of similar numerical ceilings on cruisers and smaller surface vessels was highly unlikely. Given their potential to incorporate elements of the dreadnought revolution—particularly greater firepower, range, and speed—within smaller, more cost-effective vessels cruiser could perform a wide variety of missions beyond participation in fleet operations during wartime.[121] For the smaller states, these vessels would provide some room for growth in their capabilities without the requisite expenditures demanded by the battleship. As the French delegate, Mr. Briand argued soon after the agreement to accept the 1.75 ratio on capital ships, further extension of limitations to other classes of vessel could not be accepted:

> But so far as defensive ships are concerned (light cruisers, torpedo boats and submarines), it would be impossible for the French Government, without running counter to the vote of the Chambers, to accept reductions corresponding to those for capital ships subject to this formal reservation which you will certainly understand.[122]

Ultimately, a wide variety of new cruisers were produced in the period after Washington. Perhaps most importantly, the cruiser issue would become a major issue of contention between the United States and Great Britain.

Though Great Britain had strongly supported the abolition of submarines, there was little support for this initiative. Given the losses that had been visited upon British shipping by the German U-boat campaign, it was not surprising that the official position of the British delegation would favor abolition. However, other states viewed the submarine as a legitimate vessel for a variety of missions, and it was also attractive to the lesser powers as a cost-effective countermeasure to larger naval powers. Additionally, there was a prevailing opinion that submarines were essentially defensive in nature. The German utilization of submarines may have been repugnant in its indiscriminate targeting of unarmed merchant ships, but such moral considerations should not overwhelm other perceived attributes of this class of vessel.[123] Italy joined its most likely naval rival in supporting the utility of the submarine as a primarily defensive asset. The Italian naval experts thought that the submarine was still an indispensable weapon for the defense of the Italian coasts, which had a very great extent and along which some of the largest cities, the principle railways, and a number of the most important industrial establishments were situated.[124]

Similarly the Imperial Japanese delegation also saw submarines as particularly versatile and capable of contributing to a wide variety of important missions. They were less persuaded by the arguments derived from the German experience.[125] For the United States, the development of the submarine was viewed as important. Though they could appreciate the British view on submarines after the unlimited submarine warfare that had been unleashed upon British commercial and merchant shipping, the submarine offered capabilities that the American delegation was unwilling to forego. Given the geography of the United States and the need to defend the nation from threats across two different large oceans, submarines would seem to be particularly useful in scouting and defensive roles.[126] Ultimately a fairly weak agreement offered by U.S. delegate Elihu Root limited the utilization of submarines in the event of hostilities to legitimate naval activities to avoid the more deleterious actions seen in the past war.[127] Specifically, the use of submarines against unarmed merchant shipping was prohibited.[128]

PREVAILING VIEWS OF SURVEILLANCE

Given that massive scale of battleship building programs and the length of time, requisite infrastructure, and industrial and technological capacity to produce these massive ships, there was a presumption throughout the Washington Conference period that naval disarmament could be effectively monitored and verified through existing free press and media accounts as well as the work of intelligence officials and military and naval attaches abroad. During the course of the "dreadnought revolution," even the leading edges of technical innovation seemed to diffuse quite rapidly in large part due to relative transparency of the respective building programs across the globe. In fact this openness is captured in Secretary Hughes's famous and dramatic opening statement, which described in detail the various programs of the participating states—to the vessel—in order to set the stage for his call for significant reductions. Therefore, if the negotiations resulted in an agreement to limit number of capital ships (and most classes of surface vessels for that matter) then the resulting treaty would be effectively "self-policing." There was little fear that a state could manage to illicitly build a breakout naval capability.

As will be discussed in the next chapter, the challenge of monitoring the development of military and naval airpower was much more difficult. Straightforwardly, aircraft and their major components could be built in relatively confined areas that would make effective inherently surveillance difficult. Second, the emerging civilian use of aviation presented a major challenge for any attempts to limit the military applications of aviation. The development of heavier-than-air platforms for civilian or commercial purposes could clearly have major implications for military operations over time. Critical improvements in speed, performance, range, or payloads could easily be translated into military aircraft. The nominal limitations on carriers would allow for continued experimentation without attempting to restrain what was seen as technological progress.

MOVING FORWARD: BUILDING UPON SUCCESS

Finally, Secretary Hughes was sensitive to the fact that technological change would likely make revisiting the issues of arms limitation a logical requirement. Thus he offered the idea of the conference as a first step, rather than a final word, for the cooperative endeavor of arms limitation.[129] The minimal limitations on aircraft and carriers, and the inability to limit submarines and smaller surface vessels did not imply that such limitations could not be developed in the future. Hughes saw the importance of revisiting the arms limitation issue, to build upon the accomplishments achieved at the Washington Conference. Similarly, Hughes was cognizant of the fact that political realities could change over time as well, and that new challenges and opportunities could arise. Thus a follow-on conference seemed like a logical component of the settlement. The Five-Power Treaty was a signal accomplishment. The limitation of capital ships and the proscription to forego the development of fortifications in the western Pacific combined to reshape the security environment in the region. The Four-Power Treaty replaced the troubling Anglo-Japanese alliance with a more consultative process to manage and resolve future disputes. Together, the improvement in the Pacific security environment also allowed for the reconciliation of diplomatic problems associated with China through the Nine-Power Treaty.

CONCLUSIONS: ASSESSING MET AND ALTERNATIVE HYPOTHESES

The Washington Naval Conference produced a series of treaties that sought to provide a comprehensive solution to the problems of competitive naval building and political conflict in the Pacific. Central to the success of the conference was conclusion of the Five-Power Treaty limiting the capital ship inventories of the United States, Great Britain, Japan, France, and Italy. This agreement effectively ended what would have been a costly trilateral arms race, dampened a growing rivalry between the United States and Britain and broke an increasingly intense security dilemma in the Pacific. The agreed ratios on capital ships, those deemed to be most important for the actual prosecution of a naval war, and the agreement to forego the fortification of insular possessions transformed the security environment in the Pacific. By cooperatively restraining their ability to project power and develop forward bases from which to support that power, the spiral of insecurity that had accelerated since the end of the Great War was halted, if not reversed. Combined with the Four-Power Treaty, which replaced the Anglo-Japanese alliance, and the Nine-Power Treaty, which (albeit incompletely) reconciled the various difficulties associated with a troubled China, the Five-Power Treaty created a new status quo in the Pacific that enhanced the security of contracting powers and thus decreased the probability of war. In total, the Washington Conference was a significant diplomatic achievement and represents a successful case of security cooperation.

However, it was not perfect. Within twenty years, there would be another war and the United States and Japan would engage in the very type of total naval conflict that the participants at Washington sought to avoid. This unfortunate fact tends to obscure what was accomplished at Washington. There are limits to what any diplomatic agreement can deliver. The continuing instability of China would frustrate American aspirations for the Open Door and provide opportunity for Japanese expansion. There is little that the parties at Washington could do to solve the internal turmoil in China. Even in agreeing to the Nine-Power Treaty and acknowledging the sovereignty and territorial integrity of China, Japan maintained its "special interests" in the mainland and defined the Pacific Rim as within its own sphere of influence. The United States implicitly acknowledged this reality, but continued to view China through the lens of the Open Door policy and thus maintained an inherent commitment there. The mutually contradictory policies would ultimately lead to war in the Pacific in 1941. Without a dramatic redefinition of interests by one or both powers it seems that war may have been to some degree inevitable. Nevertheless, the Washington Conference constituted a cooperative diplomatic endeavor that allowed the contracting parties to avert a war in the 1920s. In doing so, it significantly improved the security of the parties, thus leaving open the possibility of further cooperation. The fact that subsequent attempts to build upon the Washington Conference Treaties failed in the face of global turmoil do not negate their significance.

As discussed above, the critical factor that allowed for a successful cooperative outcome in the Five-Power Treaty was the collective assessment of the dreadnought battleship.[130] While the dreadnought certainly represented a revolutionary application when it was first launched, by the time of the Washington Conference this was no longer the case. By the end of the Great War, its capabilities and (more importantly) its limitations were apparent. Technological advance had provided glimpses of new applications that offered the potential to transform naval warfare with troubling implications for the big-gun battleship. Aviation and subsurface vessels could provide dramatic new capabilities for states that invested in their development. The dreadnought had a relatively limited but potentially important role in the post-War navy; it was only truly useful in large-scale fleet engagements. Thus its utility was relegated to actual war. Smaller surface vessels that incorporated advances in gunnery and propulsion were far more versatile and could contribute to a variety of roles and missions other than war. Factoring in the costs of building larger vessels with only incremental improvements in capabilities in the face of new (possibly decisive) applications and we can understand how leaders came to see them as sensible targets for limitation. This is not to say that they were not significant or to imply that their limitation was "easy" for naval powers to accept. To the contrary, despite the arguments of airpower and submarine warfare advocates within the respective military establishments of the contracting states, the battleship was still considered the central application, the "backbone," of any navy. They were deemed essential for the effective prosecution of naval war. Preparedness dictated their continued service, but their utility was limited and their longer-term future was unclear. Therefore, the delegates to the Washington Conference and their respective political and military leaderships could agree to limitations on these assets once certain agreed ratios could be negotiated. These ratios would maintain the numbers of ships deemed necessary to provide security but would also make offensive operations difficult. The 5:5:3 ratio combined with the agreement to prohibit the fortification of insular possessions provided a formula that would enhance the security of the parties and create an acceptable status quo in the Pacific.[131]

The Washington Naval Conference provides strong evidence for MET's general propositions as well as several specific hypotheses. The story of the negotiation over the dreadnought battleship clearly affirms the second and third general hypotheses. The United States, Great Britain, and Japan viewed capital ships as known applications based on mature technologies that would only have limited incremental security benefits in the short term. The size and scope of shipbuilding programs and requisite infrastructure similarly made surveillance of any agreement relatively straightforward. Moreover, despite the uncertain global security environment and the presence of a security dilemma in the Pacific, American, British, and Japanese leaders and their delegates were able to use agreements based on the restraint of these vessels to cooperatively improve their security and decrease the probability of a conflict in the short term. Unsurprisingly, as the second "laggard" hypothesis would predict, Japan proved the hardest bargainer of the three major powers, demanding the completion and deployment of the *Mutsu* and thus achieving a perceived numerical level deemed necessary for security.

Conversely, the inability to construct effective limits on naval aviation and aircraft carriers, submarines, and smaller surface vessels like cruisers, support the first general hypothesis of MET, as well as the specific non-cooperative hypotheses ($\mathbf{H}_{\mathbf{Lead2}}$ and $\mathbf{H}_{\mathbf{Lag1}}$) given the perceived value of these emerging technologies and the difficulties associated with verification and compliance. Arguments over the legitimacy of their employment during the war notwithstanding, there were consensus that the submarine had dramatic potential beyond its previous contributions. Moreover, it was viewed as a relatively cost-effective platform that could, in the shorter term, significantly contribute to coastal defense. Precisely because of this perceived value and the potential for a more decisive role in the future, the only agreement that could be reached was the (relatively weak) treaty setting out rules for submarine warfare. Similarly, no agreement was reached on aircraft. Given the rapid progress that had been made in aviation, there seemed to be little practical means to effectively limit military exploitation of this breakthrough technology. The agreed limitation on aircraft carriers was a half-measure that would provide for the capacity needed to continue experimentation and development of airpower, but would constrain the quantitative advantage that any power could gain. Finally, a consistent criticism of the Washington Conference was its failure to formalize limits on cruisers. With limitations of battleships and the diffusion of naval technology, the prevailing view that smaller vessels with greater range and mobility could execute a variety of missions necessary for naval defense, even if their relative value in wartime was as auxiliary to battleships in the context of fleet maneuvers. The failure to limit cruiser numbers did have the unintended effect of leading to an expansion of this class of vessel by the major powers and became a source of contention between the United States and Great Britain.[132] Moreover, technological trends eventually moved designers toward smaller ships with heavier firepower and armor culminating in the "pocket battleship." However, this reflects the inherent difficulty of developing comprehensive arms control limitations over a longer term and is more a failure of states to cooperatively adapt to technological change than a failure of the Washington Conference.

The Five-Power Treaty and the diplomatic negotiations of the Washington Conference clearly support the predictions of MET. In this case the stable nature of battleship technology provided a necessary condition that allowed for successful bargaining between the representatives of the contracting powers. While continuing to be a central component of an effective navy, the dreadnought was a known quantity. Possessing a certain quantity of these assets would allow the United States, British, and Japanese governments to defend their vital interests in the event of a war and constrained the offensive capabilities they could project. This technological reality provided the foundation for a credible bargain to be struck and thus played a central role in precipitating the cooperative development of a new status quo in the Pacific.

The alternative hypotheses perform poorly here. The overall success of the Washington Conference and Five-Power Treaty disconfirms offensive realist predictions. Similarly, given the relatively intense security dilemma in the Western Pacific, it seemed unlikely that defensive realism would have predicted cooperation. Perhaps more importantly, even if the nature of naval (versus continental) rivalry may have contributed to the ability to cooperatively restrain the most threatening offensive weapons (capital ships, as well as potential fortified bases near Japan), the variation in cooperation that we see within the case cannot be explained by a traditional defensive realist account. The lack of any formal oversight or verification organization underscores the limitations of an institutionalist contribution to the eventual cooperative outcome. Moreover, the conference was a relatively spontaneous event, taking place less than a year into the Harding administration. The United States was not a member of the League of Nations, and while the concept of disarmament had gained salience during the League negotiations and was a goal of the League charter, the Washington Conference took place outside of the League auspices. The delegations provided their own technical experts and the "preparatory work" was primarily done by Secretary Hughes and Theodore Roosevelt Jr. Surveillance and compliance issues were minimal. It was generally assumed that large-scale violations would be readily discovered. Thus, the cooperation achieved at Washington was multilateral, but cannot be categorized as institutional.

Finally, while domestic economic pressures within all three powers did seem to influence the thinking of the respective leaders and delegates, a failure to consummate the treaty may have indeed forced these countries to reluctantly devote scare resources to the building programs that had been devised prior to the conference. The eventual emergence of a "cruiser race" that would undermine U.S. and British relations later in the 1920s seems to further question the applicability of a domestic-budgetary (rather than security) driven hypothesis.[133] Similarly, shared values or worldviews played a negligible role at Washington. Aside from possessing security-seeking motives, the representatives of the powers had real disagreements about the respective positions of their states and the proper levels of naval power that was appropriate.[134]

NOTES

1. On the history of the Washington Conference, see for example: William Reynolds Braisted, *The United States Navy in the Pacific, 1900–1922* (Austin: University of Texas Press 1971); Thomas H. Buckley, *The United States and the Washington Conference, 1921–1922* (Knoxville: University of Tennessee Press, 1970), 101–6; Raymond Leslie Buell, *The Washington Conference* (New York: D. Appleton, 1922); Roger Dingman, *Power in the Pacific: The Origin of Naval Arms Limitation, 1914–1922* (Chicago: University of Chicago Press, 1976), 58–60; Erik Goldstein and John Maurer, eds., *The Washington Conference, 1921–22: Naval Rivalry, East Asia Stability and the Road to Pearl Harbor* (Portland: Frank Cass, 1994); Yamato Ichihashi, *The Washington Conference and after, a Historical Survey* (Stanford, CA: Stanford University Press, 1928); Mary Klachko, "Anglo-America Naval Competition, 1918–1922" (PhD dissertation, Columbia University, 1962); Bejamin H. Williams, *The United States and Disarmament* (New York: Whittlesey House, 1931).

2. On the "war scare," and Japanese policy more generally, see, Sadao Asada, *From Mahan to Pearl Harbor: The Imperial Japanese Navy and the United States* (Annapolis, MD: Naval Institute Press, 2006), 51.

3. Emily O. Goldman, *Sunken Treaties: Naval Arms Control between the Wars* (University Park: Pennsylvania State University Press, 1994); Malcolm H. Murfett, "Look Back in Anger: The Western Powers and the Washington Conference of 1921–1922," in *Arms Limitations and Disarmament: Restraints on War, 1899–1939*, ed. B. J. C. McKercher (Westport, CT: Praeger, 1992).

4. On the dreadnought battleship, see, for example, Raymond C. Gamble, "Decline of the Dreadnought: Britain and the Washington Naval Conference, 1921–1922" (PhD dissertation, University of Massachusetts, 1993); Nicholas A. Lambert, *Sir John Fisher's Naval Revolution* (Columbia: University of South Carolina Press, 2002); Robert K. Massie, *Castles of Steel, Britain, Germany, and the Winning of the Great War at Sea* (New York: Ballantine Books, 2003); Robert K. Massie, *Dreadnought: Britain, Germany, and the Coming of the Great War* (New York: Random House, 1991).

5. See, for example, O'Connell, *Sacred Vessels*, 240–47.

6. Ibid., 253–60.

7. Paul M. Kennedy, *The Rise and Fall of British Naval Mastery* (London: Ashfield Press, 1983); Stephen Roskill, *Naval Policy between the Wars, Volume 1: The Period of Anglo-American Antagonism, 1919–1929* (New York: Walker, 1968); Harold Sprout and Margaret Sprout, *Toward a New Order of Sea Power: American Naval Policy and the World Scene, 1918–1922* (New York: Greenwood Press, 1976).

8. B. J. C. McKercher, ed., *Anglo-American Relations in the 1920s: The Struggle for Supremacy* (Basingstoke: Macmillan, 1991).

9. See, for example, Klachko, "Anglo-America Naval Competition, 1918–1922," 118–34.

10. On the U.S. building programs, see Sprout and Sprout, *The Rise of American Naval Power, 1776–1918*, 344–46.

11. Erik Goldstein, "The Evolution of British Diplomatic Strategy for the Washington Conference," in *The Washington Conference, 1921–22: Naval Rivalry, East Asia Stability and the Road to Pearl Harbor*, ed. Erik Goldstein and John Maurer (Portland: Frank Cass, 1994), 10–12.

12. Dingman, *Power in the Pacific*, 67–88.

13. Klachko, "Anglo-America Naval Competition, 1918–1922," 53–90; Roskill, *Naval Policy between the Wars*, 204–34.

14. Buckley, *The United States and the Washington Conference, 1921–1922*, 20–23.

15. On U.S. naval power during the period see, for example, Kennth J. Hagan, *This People's Navy: The Making of American Seapower* (New York: Touchstone, 1991); Robert L. O'Connell, *Sacred Vessels: The Cult of the Battleship and the Rise of the U.S. Navy* (New York: Oxford University Press, 1991); Harold Sprout and Margaret Sprout, *The Rise of American Naval Power, 1776–1918* (Princeton, NJ: Princeton University Press, 1946).

16. Buckley, *The United States and the Washington Conference, 1921–1925*, 25.

17. Asada, *From Mahan to Pearl Harbor: The Imperial Japanese Navy and the United States*, 58–60; Sadao Asada, "Japan's 'Special Interests' and the Washington Conference," *American Historical Review* 61, no. 1 (1961); Braisted, *The United States Navy in the Pacific, 1900–1922*, 535–48.

18. Walter LaFeber, *The Clash: A History of U.S.-Japan Relations* (New York: W. W. Norton, 1997), 106–10.

19. Asada, *From Mahan to Pearl Harbor: The Imperial Japanese Navy and the United States*, 51.

20. Roskill, *Naval Policy between the Wars*, 131–81. George Kennan, *The Decision to Intervene* (New York: Atheneum, 1967).

21. David C. Evans and Mark R. Peattie, *Kaigun: Strategy, Tactics, and Technology in the Imperial Japanese Navy 1887–1941* (Annapolis, MD: Naval Institute Press, 1997), 187–91; Edward S. Miller, *War Plan Orange: The U.S. Strategy to Defeat Japan, 1897–1945* (Annapolis, MD: Naval Institute Press, 1991), 86–99; Steven T. Ross, *American War Plans 1890–1939* (London: Frank Cass, 2002).

22. Asada, *From Mahan to Pearl Harbor: The Imperial Japanese Navy and the United States*, 58–60.

23. Evans and Peattie, *Kaigun*, 150–67.

24. See, for example, Sir B. Alston respecting suggestions for an Anglo-Saxon Policy for the Far East," August 1, 1920 in *Documents on British Foreign Policy, 1919–1939: Volume XIV, Far Eastern Affairs, April 1920–February 1922*, ed. Great Britain Foreign Office (Oxford: Her Majesty's Stationary Office, 1966), 81–86; Michael Graham Fry, "The Pacific Dominions and the Washington Conference, 1921–22," in *The Washington Conference, 1921–22: Naval Rivalry, East Asian Stability and the Road to Pearl Harbor*, ed. Erik Goldstein and John Maurer (Essex: Frank Cass, 1994), 64.

25. Braisted, *The United States Navy in the Pacific, 1900–1922*, 549–64; Roskill, *Naval Policy between the Wars*, 278.

26. Michael Graham Fry, "The Pacific Dominions and the Washington Conference, 1921–22," in *The Washington Conference, 1921–22: Naval Rivalry, East Asian Stability and the Road to Pearl Harbor*, ed. Erik Goldstein and John Maurer (Essex: Frank Cass, 1994), 64.

27. Goldstein, "The Evolution of British Diplomatic Strategy for the Washington Conference," 11.

28. Dingman, *Power in the Pacific*, 122–28; Evans and Peattie, *Kaigun*, 166–75; Roskill, *Naval Policy between the Wars*, 221–24.

29. Evans and Peattie, *Kaigun*, 152–67; Archer Jones and Andrew J. Keogh, "The Dreadnought Revolution: Another Look," *Military Affairs* 49, no. 3 (1985); Lambert, *Sir John Fisher's Naval Revolution.*

30. On Jutland and its impact, see Geoffrey Bennett, *The Battle of Jutland* (Hertfordshire: Wordsworth Editions, 1999); George Bonney, *The Battle of Jutland 1916* (London: Sutton Publishing, 2006); John Campbell, *Jutland: An Analysis of the Fighting* (London: Lyons Press, 1998); David MacGregor, "The Use, Misuse, and Non-Use of History: The Royal Navy and the Operational Lessons of the First World War," *Journal of Military History* 56, no. 4 (1992); Arthur J. Marder, "The Influence of History on Sea Power: The Royal Navy and the Lessons of 1914–1918," *Pacific Historical Review* 41, no. 4 (1972); Arthur J. Marder, *Jutland and after (May 1916–December 1916)*, vol. 3, *From the Dreadnought to Scapa Flow: The Royal Navy in the Fisher Era, 1904–1919* (London: Oxford University Press, 1966).

31. An excellent retelling of the fighting at Jutland can also be found in O'Connell, *Sacred Vessels*, 153–89.

32. Evans and Peattie, *Kaigun*, 167.

33. Daniel Allen Butler, *Distant Victory: The Battle of Jutland and the Allied Triumph in the First World War* (London: Praeger, 2006).

34. Dingman, *Power in the Pacific*, 21–23.

35. Committee on Foreign Affairs, U.S. House, *Disarmament*, 66th Congress, 3rd Sess., 1921, 67.

36. Dingman, *Power in the Pacific*, 106–21; Gamble, "Decline of the Dreadnought: Britain and the Washington Naval Conference, 1921–1922," 196–241; Roskill, *Naval Policy between the Wars*, 223–25.

37. Roskill, *Naval Policy between the Wars*, 221.

38. Gamble, "Decline of the Dreadnought: Britain and the Washington Naval Conference, 1921–1922," 198.

39. Ibid.

40. Ibid, 221–41.

41. Braisted, *The United States Navy in the Pacific, 1900–1922*, 477–99; Dingman, *Power in the Pacific*, 95–99; Sprout and Sprout, *Toward a New Order of Sea Power*, 73–76.

42. Sprout and Sprout, *Toward a New Order of Sea Power*, 75–76.

43. Committee on Naval Affairs, U.S. House of Representatives, *Disarmament*, 66th Cong., 3rd Sess., 1921, 577.

44. Sprout and Sprout, *The Rise of American Naval Power, 1776–1918*, 345.

45. Evans and Peattie, *Kaigun*, 170–71.

46. Evans and Peattie, *Kaigun*, 170–71.

47. Ibid., 171.

48. On the Japanese building programs, see Evans and Peattie, *Kaigun*, 150–51; J. Charles Schencking, *Making Waves: Politics, Propaganda, and the Emergence of the Imperial Japanese Navy, 1868–1922* (Stanford, CA: Stanford University Press, 2005), 211–21.

49. Hagan, *This People's Navy: The Making of American Seapower*, 269–72; O'Connell, *Sacred Vessels*, 240–47. While O'Connell views the continued push for dreadnoughts as "wasteful, dangerous and largely ridiculous," he downplays the forces within and outside of the Navy leadership that were pushing for change and making progress. Faced with a short-term Japanese threat, it is not (and should not be) surprising that the General Board and the CNO would continue to maximize its dreadnought force.

50. See for example, Williamson Murray, "Strategic Bombing: The British, American, and German Experiences," in *Military Innovation in the Interwar Period*, ed. Williamson Murray and Allan R. Millett (Cambridge: Cambridge University Press, 1996); Geoffrey Till, "Adopting the Aircraft Carrier: The British, American, and Japanese Case Studies," in *Military Innovation in the Interwar Period*, ed. Williamson Murray and Allan R. Millett (Cambridge: Cambridge University Press, 1996).

51. On Mitchell, see Alfred F. Hurley, *Billy Mitchell: Crusader for Air Power* (Bloomington: Indiana University Press, 2006); Douglas C. Waller, *A Question of Loyalty: Gen. Billy Mitchell and the Courtmartial That Gripped the Nations* (New York: HarperCollins, 2004).
52. Ibid.
53. Ibid., 23.
54. Ibid., 33–34.
55. Ibid., 32–33.
56. Ibid., 226.
57. Ibid., 226.
58. Roskill, *Naval Policy between the Wars*, 264.
59. Gamble, "Decline of the Dreadnought," 237.
60. Ibid., 217–28.
61. See, for example, Williamson Murray, "Strategic Bombing: The British, American, and German Experiences," in *Military Innovation in the Interwar Period*, ed. Williamson Murray and Allan R. Millett (Cambridge: Cambridge University Press, 1996); Geoffrey Till, "Adopting the Aircraft Carrier: The British, American, and Japanese Case Studies," in *Military Innovation in the Interwar Period*, ed. Williamson Murray and Allan R. Millett (Cambridge: Cambridge University Press, 1996).
62. Ibid., 32.
63. Ibid., 33–34.
64. Ibid., 32–33.
65. "Testimony of Admiral William Sims," in *Hearings on Naval Policy*, 656.
66. Ibid., 657–58.
67. Ibid., 660.
68. Ibid., 660–61.
69. Ibid., 661.
70. Evans and Peattie, *Kaigun*, 180.
71. Mark R. Peattie, *Sunburst: The Rise of Japanese Naval Air Power, 1909–1941* (Annapolis, MD: Naval Institute Press, 2001), 9.
72. Ibid., 10.
73. Ibid., 13.
74. Evans and Peattie, *Kaigun*, 150–67.
75. Schencking, *Making Waves*, 214.
76. Peattie, *Sunburst*, 100.
77. See "Testimony of Admiral Robert E. Coontz, Chief of Naval Operations," in Peattie, *Sunburst*, 762.
78. Ibid., 686.
79. On the cruiser question, see Roskill, *Naval Policy between the Wars*, 326–27.
80. Sprout and Sprout, *Toward a New Order of Sea Power*, 208–9.
81. Buckley, *The United States and the Washington Conference*, 118–19.
82. Notes by Admiralty Section of the British Delegation, Washington, November 13, 1921, "Auxiliary Combatant Craft," from *Anglo-American Naval Relations, 1919–1939* Michael Simpson, ed. (Surrey: Ashate for the Naval Records Society, 2010), 43.
83. Ibid, 42.
84. On cruiser development after Washington, see John Jordan, *Warships after Washington: The Development of the Five Major Fleet 1920–1923* (Barnsley: Seaforth Publishing, 2011), especially 108–52. Also see John H. Maurer and Christopher M. Bell, *At the Crossroads between Peace and War: The London Naval Conference of 1930* (Annapolis, MD: Naval Institute Press, 2014).
85. Despite this importance, submarine programs were often underfunded in the context of budget austerity. See Holger H. Herwig, "Innovation Ignored: The Submarine Problem, Germany, Britain, and the United States, 1919–1939" in Williamson Murray and Allan R. Millett, ed., *Military Innovation in the Interwar Period* (Cambridge: Cambridge University Press, 1996), 252–60.
86. Herwig, "The Battlefleet Revolution, 188–1914" in Knox and Murray, eds., *The Dynamics of Military Revolution 1300–2050*, 115.

87. Sprout and Sprout, *Toward a New Order of Sea Power*, 2, 79.
88. *Conference on the Limitation of Armaments*, 275.
89. See, for example, Roskill, *Naval Policy between the Wars*, 326–28.
90. Buckley, *The United States and the Washington Conference, 1921–1922*, 68.
91. Ibid., 70.
92. Ibid., 70–73; *Conference on the Limitation of Armamen*t, 41–50.
93. *Conference on the Limitation of Armament*, 46.
94. Ibid., 47.
95. Buckley, *The United States and the Washington Conference, 1921–1922*, 49–56; *Conference on the Limitation of Armament*, 56–63; Dingman, *Power in the Pacific*, 155–59; Roskill, *Naval Policy between the Wars*, 308–12; Sprout and Sprout, *Toward a New Order of Sea Power*, 141–48.
96. Buckley, *The United States and the Washington Conference, 1921–1922*, 73–74.
97. Ibid., 73.
98. *Conference on the Limitation of Armament*, 48.
99. Ibid., 50–55.
100. The minutes of these meetings are reprinted in *Conference on the Limitation of Armament*, 237–440.
101. Ibid., 254–57.
102. *Frus, 1922,* I, 135.
103. Ibid., 65.
104. Ibid., 67–68.
105. Ibid., 76–77.
106. Ibid., 77.
107. Ibid., 78.
108. Ibid., 74–75.
109. Ibid., 81.
110. Ibid., 83.
111. Ibid., 84.
112. Ibid., 86.
113. Ibid., 89.
114. Ibid., 90.
115. Ibid., 128.
116. Ibid., 136.
117. *Frus, 1922,* I, 357.
118. Sprout and Sprout, *Toward a New Order of Sea Pow*er, 231–36.
119. *Frus, 1922,* I, 388–410.
120. Sprout and Sprout, *Toward a New Order of Sea Power*, 229.
121. For an excellent discussion of the negotiations on other classes of weapons, see Buckley, *The United States and the Washington Conference, 1921–1922*, 113–26.
122. Ibid., 136.
123. Ibid., 270.
124. Ibid., 271.
125. Ibid.
126. Ibid., 276.
127. *Conference on the Limitation of Armament*, 325–45.
128. The section on legitimate and illegitimate submarine warfare can be found in *Conference on the Limitation of Armament*, 813–17.
129. *Frus, 1922,* I, 259.
130. Goldman, *Sunken Treaties*, 80.
131. Ibid., 106–7.
132. See, for example, Allen W. Dulles, "The Threat of Anglo-American Naval Rivalry," *Foreign Affairs* 7, no. 2 (January 1928): 173–82; B. J. C. McKecher, "A Certain Irritation: The White House, the State Department, and the Desire for a Naval Settlement with Great Britain, 1927–1930," *Diplomatic History* 31, no. 5 (November 2007): 829–63.

133. Dingman, *Power in the Pacific*, 157; Maurer, "Arms Control and the Washington Conference," 283–86.
134. Roskill, *Naval Policy between the Wars*, 544–66.

Chapter 3

The World Disarmament Conference

Beyond the development of the atomic bomb and the crossing of the nuclear threshold at the end of the Second World War, perhaps no period of time witnessed greater technological change with truly dramatic implications for military innovation and international security than the Interwar Period (1919–1939).[1] During this time, amidst the emergence of a range of potentially decisive new weapons, leading states engaged in a substantive attempt to achieve comprehensive disarmament agreements in concert with cooperative efforts to forge a sustainable peace. Understanding the strongly defensive character of the security environment of the Great War, and viewing the emergence of new military applications that could overturn it, leaders attempted to perpetuate that environment. However, despite their attempts to achieve a cooperative outcome, the great powers effectively chose instead to allow for the development and acquisition of new military capabilities that would overcome the existing defensive security environment and could make the next war short and decisive.[2] This presents an interesting empirical puzzle. If leaders understood the defense dominant nature of the existing system, and could also determine the types of weapons that may facilitate offensive operations (or at least make them viable), why would agreement among security-seeking states prove impossible? Certainly by 1933, when the World Disarmament Conference at Geneva ultimately collapsed, the likelihood of successful cooperation was minimal. However, given that the Preparatory Commission charged with development of a draft treaty began its work in the mid-1920s under a more favorable political environment when decisive innovations seemed much further away, the answer seems far less clear.

This chapter attempts to explain this puzzle. Like the origin of the Second World War, the failure of the Geneva Disarmament Conference is somewhat over-determined. An offensive realist explanation is straightforward and compelling. The uncertainty created by the highly dynamic nature of technology forces states to acknowledge that no defense dominant world can ever be permanent and so arms control would be doomed to failure. However, a fair reading of the history and empirical evidence of the case does not support this overly simplistic story. Statesmen and their military advisors seemed to be driven by motives much more accurately depicted by defensive realism. However this case is ultimately a difficult one for that approach. If leaders understood the defense dominant nature of the existing security environment and could also determine the types of weapons that may facilitate offensive operations, why would rational security seeking states fail so miserably to achieve an effective agreement? The great powers attempted to negotiate a comprehensive agreement to maintain the defense-dominant system but failed, and therefore the 1932 World Disarmament Conference provides an important case for MET.[3] Against the backdrop of strong public pressures favoring disarmament, a global economic decline that forced states to question their respective budgetary commitments to military forces, and foreboding signs of reaction and fascism, the leading statesmen of the day failed to come to a meaningful agreement. Moreover, their discussions, which focused primarily on *qualitative disarmament*, creating prohibitions on those weapons deemed most aggressive in character and thus most capable of overturning the defense dominant system, were remarkably prescient, foreshadowing the type of war that would be visited on Poland, France, and Russia within ten years.[4] Participating nations had strong incentives to curtail spending on new "aggressive" armaments, and thus maintain their relatively security given the defensive nature of the system. However, the participating states failed to reach an agreement, which set the stage for the rearmament of Germany and the eventual war to come.

As MET would predict, the Geneva Conference provides us with a clear case in which statesmen and their advisors had sound understandings of both the existing security environment and the likely military impact of emerging technologies, yet they were unwilling and/or unable to agree on limitations precisely because of the uncertainty inherent in the dynamic nature of technology and an inability to develop a regime that could sufficiently verify compliance with an agreement. What makes the Geneva conference such a useful case is the fact that to a significant extent, the representatives did actually come to relative agreement as to what constituted aggressive armaments and therefore understood those that were most destabilizing to the existing system.[5] However, even with this relative consensus, difficult questions concerning the abolition of these new systems and monitoring their removal from national arsenals proved impossible to resolve. Ultimately, two plans centering on "abolition" or "internationalization" emerged but neither could satisfactorily address the fears and concerns of key states, precluding a negotiated settlement.

The remainder of this chapter is divided into six sections. The first section will consider the post-Great War security environment. The costs in blood and treasure of the First World War remained central to the considerations of many interwar leaders. Given the existing defense dominant system, it was unlikely that any state could truly launch an aggressive war again. However, technological change and the military applications of technologies that emerged at the end of the Great War and matured during the interwar period provided leaders with expectations that new applications like tanks, bombing, and attack aircraft, and doctrinal and strategic innovations to employ them, that could dramatically improve battlefield coordination could dramatically overturn the defense dominant system in the years to come. These expectations will be examined in the second section, as will the perceived challenges associated with effectively monitoring and verifying these decisive new armaments. The third section will consider the emergence of quantitative disarmament as the means to addressing the challenges confronting Europe's leaders. The fourth section will revisit the history of the World Disarmament Conference in Geneva. Focusing primarily on the key roles of Great Britain and France, a picture emerges of statesmen with a clear understanding of both the existing security environment and the likely impact of new military innovations, but unable to forge a bargain that could limit the most dangerous systems with any real certainty. While Great Britain's resistance to assure French security hindered deeper cooperation, the failure of the conference would hinge on the inability to reconcile competing plans: "abolition" forwarded by the British and "Internationalization," pushed vigorously by France. To opponents, the former simply avoided the questions of verification and compliance, leaving far too great an opportunity for states to cheat and rearm covertly while the latter was predicated on the development and empowerment of a powerful multilateral entity that would make unacceptable intrusions on state sovereignty. The fifth section will consider the relative influence of political and technological forces in determining the outcome of the conference. A brief conclusion will consider the respective explanatory power of hypotheses generated by MET against other alternative theoretical approaches.

THE POST-GREAT WAR SECURITY ENVIRONMENT

Emerging from the unprecedented carnage of the First World War, where the impact of trench warfare, barbed wire, and machine guns made massed infantry assaults prohibitively costly; statesmen came to recognize what can be considered the *defense dominance* of the battlefield.[6] Given existing technology and geography, offensive operations were extremely difficult to execute successfully, while defensive measures were fairly easy to employ effectively.[7] Contrary to flawed perceptions held by a consensus of military leaders across Europe, the war would be neither short nor decisive. The combined impact of infantry firepower: repeating rifles, machine guns, artillery, trenches, barbed wire, and fortifications had created an advantage for defenders.[8] Offensive operations and any concept of maneuver broke down in the face of trench warfare of the Western Front.[9] The results of the "cult of the offensive" were staggering casualty figures, depletion of national resources, and a brutal war of attrition that placed all nations under severe pressure for the duration of the conflict.[10] For example, the fact that the Schlieffen Plan almost succeeded is not proof that defense was not dominant—rather it shows the great lengths that troops using a strategy of maneuver had to go in order to succeed, reinforcing the notion that offense was more difficult.[11] Given the relatively primitive nature of logistics and communications of that era and the lack of mobility one can understand the roots of the failure, yet its ambitious strategic objective and its execution was impressive and nearly secured a rapid victory for Germany in the West.[12] Still, it remains a fair description of the World War I military environment to categorize it as defense-dominant, particularly on the Western Front.

While tactical adjustments and the utilization of emerging technologies toward the conclusion of the Great War created the possibility of offensive maneuver, the fact remains that both the famed Ludendorf offensive in 1918 and the French offensives of 1917 were devastating for the initiators.[13] The former effectively exhausted German manpower while the latter lead directly to a revolt that almost threatened to take France out of the war. While scholars have highlighted the ability of various militaries to adapt at the tactical level to better confront the challenges of trench warfare during the Great War, their respective arguments fail to discredit offense-defense theory.[14] As one scholar has written, "Ludendorf's tactical success was an operational and strategic failure."[15] Even while new applications like tanks or airplanes or novel tactics like small-group infiltration maneuvers or coordinated indirect heavy artillery strikes were introduced and had some notable effects toward the end of the war, the environment remained strongly defense dominant: large-scale offensive operations remained extremely costly and the ratio of offensive forces needed to overcome the defenses deployed far exceeded the forces needed to defend a given area.[16]

> In other words, by 1914 a cycle of weapon development had enabled the defense to gain ascendancy over the offense. In the view of this mechanical school, artillery, though responsible for limited successes, was unable, in spite of lavish use, to convert the deadlock into a war of movement. It remained for the tank, motor transport, and the aeroplane, in the closing stages of the struggle to exercise more decided tactical effects and to exploit the results of ocean blockade.[17]

As the quote implies, had the war gone on (and had tanks and airplanes been utilized in large numbers in penetrating maneuvers, for example), it may have be true that the environment could have shifted more to the offense.[18] Yet at the close of the war and immediately following the Treaty of Versailles, the world was (and was correctly perceived) as defense dominant. Offense was prohibitively costly. Trench warfare, with the combined influence of machine guns, hardened bunkers, barbed wire, and miles of entrenchments, provided states with a defensive capacity without necessarily threatening their neighbors.[19]

Given the postwar environment, where defense was dominant and defensive capabilities were distinguishable from those of the offense, what is fascinating was the eventual (and almost universal) response to this generally accepted view.[20] Reinforcing the point, these critiques underscore the fact that the offense-balance at any given time is indeed a relevant factor in the calculations of states and their militaries. Understanding the nature of the postwar security environment, which clearly seemed defense-dominant in the absence of concerted effort to develop the specific technologies that could possibly overturn the existing system (armor, mechanization, aviation), we would expect that rational, security-seeking leaders would attempt to take steps to maintain that system (or "lock it in") through cooperation.[21] Ostensibly this is what the disarmament project undertaken by the League of Nations was attempting to achieve, though their near decade-long effort would ultimately fail at Geneva in 1933.

Throughout much of the postwar period, the traditional great powers were exhausted and the impact of the war on the societies and economies of the participants cannot be overemphasized. Aside from losing generations of young men and potential productivity and economic activity associated with those losses, the financing of the war and the associated debts and calls for reparations under the Versailles Treaty would stress both victorious and defeated powers.[22] We would generally expect rational leaders to attempt to devise policies in keeping with a prevailing security environment, but given the domestic pressures faced by leaders in the aftermath of the Great War, we would expect that cooperation that could avoid costly spending on armaments so soon after a catastrophic military conflict would be viewed particularly beneficial and a desirable policy objective in terms of national security, economic reconstruction, and social welfare.

The European system in the mid-1920s after the Locarno Agreements and Germany's entry into the League was relatively stable seemed to provide the environment in which disarmament could be constructively addressed.[23] To this end, the League powers, working with both the United States and the Soviet Union, commissioned a multilateral body to examine and develop recommendations aimed at the development of a comprehensive regime that would abolish or limit emerging technologies and their perceived offensive military applications in 1926. Recognizing the importance of the task, the League of Nations succeeded in convincing the United States and the Soviet Union to contribute to the process, to develop the Draft Treaty for the World Disarmament Conference, scheduled for February 1932 in Geneva.[24] In the six years between the League resolution and the conference, political and military representatives from the participating states attempted to work out consensus on the detailed technical issues inherent in developing a comprehensive solution to the challenges of disarmament. Three commissions were tasked with examining the questions of land, air, and naval armaments respectively.[25] Technical committees investigating specific military and technological questions in turn advised these commissions. The fact that it took four years to develop a draft treaty, and two more years before the conference could be held underscores the difficulties faced in attempting to reconcile technological uncertainty in the context of self-interested state actors.[26] Nonetheless, there was considerable hope that the conference could yield positive results, and that the political representatives would be able to rectify the disagreements that plagued their military advisors.[27]

Though the conference began amidst ominous signs, most importantly the intervention of Japanese forces in Manchuria in September 1931, it is important to recognize that the political situation in Europe presented incentives for a comprehensive resolution of the disarmament question.[28] Large peace movements sent the representatives to conference with a perception of a broad public mandate for disarmament.[29] The continuing global financial crisis made military expenditures increasingly unsustainable and difficult if not impossible to reconcile in domestic political terms.[30] Similarly, while the Nazi Party had registered significant electoral gains, Germany was now a member of the League and its representatives seemed to work earnestly to achieve political equality and comprehensive disarmament.[31] Even the Soviet Union, while continuing to suffer a credibility problem because of the ongoing perception of its support for the Comintern and its political agitation, signaled a willingness to play a constructive role.[32]

THE EMERGENCE OF OFFENSIVE WEAPONS

As the 1920s wore on, important technologies introduced at the end of the Great War were maturing and were perceived to have dramatic and decisive military implications. Most importantly, these new weapons threatened to reintroduce the viability of offensive operations into military strategy. A "Revolution in Military Affairs" often perceived as "Land War Revolution" was underway, but it is perhaps more accurate to term it the "Combined Arms Revolution" which would provide innovative states with the opportunity to employ new weapons in great numbers (tanks, mobile artillery, "heavier-than-air" planes) thanks to the diffusion of industrial mass production and coordinate them through the use of emerging signals technologies, like radio.[33] These weapons had made their initial appearances toward the end of the Great War, but their limited numbers and minimal capabilities precluded their impact beyond certain discrete battles or engagements. However, as military thinkers began to consider the potential contribution of tanks and airpower, assuming some measure of command and control, the decisiveness of these new systems quickly became apparent.

> The newer instruments of war, the tank, the aeroplane, and the gasoline engine in general, appeared to afford an opportunity for developing a modernized version of a one-act war, a rapid and decisive aggression.[34]

Two weapons systems in particular would obviously dominate both the thinking of Interwar military theorists and strategists as well as the political and military leaders struggling with questions of disarmament and security. The tank and the airplane were viewed as potentially decisive weapons, an building upon the experiences of the Great War, it was not difficult to understand why.

The Tank

In a fairly straightforward way, the development of the modern tank can be understood as a response to the challenges created by trench warfare.[35] Infantry advances across contested territory in the face of withering machine gun, fire, various obstacles including barbed wired, and dug-in defensive forces had proven extremely costly on the Western front. With the advent of the tank, the combination of "Mobility, Protection and Firepower" immediately altered the existing environment.[36] Terrain could be crossed more quickly than infantry men on foot. The vehicle's armor could withstand machine gun fire and overcome fences and wire and other obstacles, and the tank's cannon could attack defender and perhaps even fortifications as time went on and deployed guns got larger. Moreover, the shock value of monstrous machines overcoming defensive fire and rolling across reinforced lines was evident from its first appearance at Cambrai.[37]

> Tanks have considerable cross-country capacity, can break down or cross wire entanglements, and combine a high degree of protection for their crews (against the fire of rifles and machine guns) with great mobility and the power to deliver a considerable volume of aimed fire while in movement.[38]

Despite their contribution to the British war effort, the large plodding machines of the Great War were very slow, often breaking down due to mechanical problems, had poor logistic and maintenance support, and ultimately played a relatively limited role. However, with the knowledge of the contribution in mind, moving forward, with mass production providing tanks in adequate numbers, better designs to improve speed, firepower, protection, and handling, and organizational and doctrinal innovation to maximize the offensive capabilities that they could provide, it was clear that the tank could be a truly decisive weapons and transform land warfare.[39] Ultimately, tanks would be found in the arsenals of all of the great powers in various sizes and shapes depending on the perceived roles and missions that they were tasked to accomplish. Some nations would prove more adept at developing and implementing concepts of armored warfare and the Second World War would bear out these differences. What was clear was that the offensive capabilities offered by tanks (and mobile artillery) were viewed as decisive across the militaries of Europe.[40]

Aviation

Perhaps in no other field had real, tangible advancements taken place in a short period of time than in aviation. From dirigibles and airships to "heavier-than-air" airplanes capable of flying increasingly long journeys carrying greater payloads incorporating greater speeds and power, manned flight had truly progressed in remarkable and potentially threatening ways. During the Great War, airplanes had been primarily limited to scouting and fire control missions, dogfights with opposing planes, and limited bombing of ground targets.[41] However, given the strides made in aviation technology and the already-apparent civilian and commercial applications that airplanes would have for communications, transport, and travel, it was clear that this application was not going away. In terms of its military applications, it seemed clear that aviation would play a central role in the next war.

> Some military men and a vast number of civilians have come to take it for granted that the next war will be fought almost exclusively in the air, and that it will be ended within a few hours or days by the destruction of the great populous centers in the territory of the less well-prepared of the combatants.[42]

As this quote implies, while aircraft could be expected to play a number of roles in a future conflict, the concept of strategic bombing seemed to grasp the imagination of many observers and experts.[43] As aircraft grew larger and larger, capable of carrying greater payloads of high explosive munitions, states may attempt to effectively avoid the loss of blood and treasure associated with a land war of competing armies by targeting the population centers of their enemies, destroying their leadership and command and control, eroding their industrial capacity and war-making capabilities, and devastating their population and cowing their people into submission. It was a particularly frightening concept, but one that would soon be put into practice in the 1940s.[44]

> Moreover, not only air enthusiasts, but sober and well-balanced writers, see in the airplane a means of striking in certain cases directly at the source of enemy power, his civilian centers and civilian will, without first having to overthrow the armed forces which protect that will.[45]

Aviation thus threatened to not just alter the nature of war, as the tank would on land, but to fundamentally transform it. States could face devastating strikes from an adversary without even being engaged in a state of hostilities. Moreover, taken to its logical end, airpower could threaten to make other investments in military forces to maintain the security of the states moot. The military implications of aviation were truly revolutionary and any means of credibly limiting or controlling offensive airpower would likely require a revolutionary solution.

THINKING ABOUT THE NEXT WAR

Throughout the interwar period, as technology seemed to offer new and decisive military applications, a generation of intellectuals emerged to consider their employment in a future conflict. Most having served in some capacity during the Great War, seemed well prepared to consider the potential solutions that new applications like tanks and aircraft could provide to the challenges that had caused so much death and devastation. In England, military J. F. C. Fuller and Basil Liddell Hart were among the first to consider the employment of large numbers of tanks in future war.[46] Their initial writings would be enthusiastically read and serve to shape the ideas of German leaders like Hans Von Seeckt and later Heinz Guederian.[47] The former was charged with maintaining and educating the severely constrained core of the professional officers allowed under the League Mandate in the early 1920s. The latter would prove to be a critically important figure in the development of Panzer tactics which would contribute to the successes of Nazi Germany's Blitzkrieg campaigns. In the Soviet Union, Marshal Mikhail Tukhachevsky, who was a dominant force in the Red Army throughout the late 1920s and up until the mid-1930s, developed the concept of Deep Battle, or deep operations theory, based upon the perceived advantages provided by the large-scale utilization of battle tanks and aircraft, which he firmly believed would play a major role in the battlefield of the future.[48] "Deep Battle" envisioned the type of powerful, focused attacks, combining independent armored attacks, working closely with tactical air power to plunge deep into enemy formations (including rear positions and reserves) that would deliver rapid and decisive victory.[49] This type of method of attack ironically became associated with Blitzkrieg and unleashed against the Soviet Union in 1941 during Operation Barbarossa.[50] Incorporating both tactics and strategy to focus at the operational level of war, Tukachevsky and his supporters were visionary, and quite far-sighted. Moreover, the under Tukachevsky's leadership of the Red Army, the Soviet Union had invested significant resources in new technologies of armor and tactical air power and developed doctrine to combine those elements into a coordinated, effective assault. As Azer Gat explains:

> Only in the Soviet Union did the proponents of "Deep Battle"—who were senior enough themselves to occupy the army's high command—formulate their ideas as an official doctrine from above for the entire army, combining mechanized and non-mechanized elements.[51]

Like Billy Mitchell in the United States, Italian General Giulio Douhet, who first commanded Italian forces in Turkey during the Great War, became a driven advocate for airpower, and contributed to the development of the concept of strategic bombing, envisioning large powerful airplanes with extensive range and defensive capabilities as being capable of delivering devastating attacks on urban and industrial centers.[52]

Ultimately the combination of airpower and fast-moving, independent armored columns, coordinated by highly professional officers possessing both requisite tactical and technical skills, would prove to be the model that would become Blitzkrieg. The capabilities provided by armor and airpower would indeed revolutionize war and restore the viability of offensive operations to the battlefield. While these weapons were necessary, the intellectual foundations provided by the military theorists of the interwar period also played a decisive role in developing the strategies, doctrines, operational concepts, and (in some cases) tactics, that made those weapons so effective. This was certainly evident after the outbreak of the Second World War when the "pendulum of military invention" had shifted in favor of the offensive.[53] What was also particularly challenging about the limitation of both armor and military aviation—the two components of aggressive warfare that most troubled military and political leaders—was the inherent difficulty of surveillance or credibly monitoring their development and production. Given the capacity of most great powers, mass production in closed factories was possible (and likely) making the challenge of surveillance a daunting one. The military implications deriving from the widespread adoption of the technology of the combustion engine proved impossible to limit with any confidence once particularly once its civilian and commercial utility became clear.

CHALLENGES TO SURVEILLANCE AND VERIFICATION

The inherent difficulty associated with monitoring aircraft was a consistent impediment to the conclusion of an agreement based upon abolition. Given the breakthrough nature of aviation, its widespread and increasing prevalence across the globe since its inception, and the its potentially dramatic implications for warfare, effective surveillance would seemingly be critical to any effective agreement.[54] The manufacture of aircraft poses a serious problem.

> Parts of airplanes may be made in modest shops in a score of different cities, and they need not even be brought together for assembly until the time when they are needed in service. After a satisfactory type has been developed, the individual machines are cast almost as much as Ford cars, and interchangeable parts may wait five years after manufacture before being put together.[55]

Moreover, as experts explained at the time of the Preparatory Commission and the conference, the mere existence of civilian or commercial air fleets in one state constitute a latent, but potentially acute threat to the security of other states.[56]

> The close similarity of certain commercial and military aircraft, and the similarity in all cases of certain parts such as the wings, constitutes another problem. The equipment for an air force may be built up under the guise of commercial activity. . . . Such a force would serve as the purpose of a "fleet in being," presenting a strategic threat which would keep the enemy from attempting the use of commercial types and other inferior from a military point of view.[57]

While more expansive and intrusive verification plans could potentially focus on the types of engines or other technical capabilities or performance characteristics that civilian planes would be allowed, other may instead monitor the numbers of pilots trained and available to states. In short, surveillance was an exceedingly difficult if not impossible task without a robust oversight capacity.[58]

> There seems to be no way of overcoming completely the possibility of evasion. A force of inspectors would have to be placed in every factory suspected of ability to build airplanes or their engines, and the work would have to be watched at every step, both in the shop and in the drafting-room.[59]

While aircraft and aviation present the most acute challenge to verification, the development of modern tanks and mobile, heavy artillery were similarly difficult. Even factories that had the capacity to mass produce farm tractors and heavy industrial equipment for civilian use could relatively easily be converted to producing these types of military applications. Certainly the outfitting of soldiers with equipment like small arms and machine guns was even more difficult to monitor with any sense of certainty. In short, not only were the weapons of this period perceived to be decisive, they were also excessively difficult to monitor, thus dampening the enthusiasm for disarmament programs based on abolition and seemingly demanding expansive and intrusive oversight mechanisms.[60]

THE MOVEMENT TOWARD QUALITATIVE DISARMAMENT

Throughout the 1920s and early 1930s, several attempts were made to limit armaments to avoid what were seen to be costly arms races that threatened fiscal situations of states leading to domestic political instability and, more importantly, to avoid escalation dynamics that could lead to another destructive war.[61] Disarmament was seen as a critical objective within the larger movement toward lasting peace embodied in the League of Nations Charter and the Locarno agreements.[62] For proponents of the League, the question of disarmament was directly tied to the resolution of underlying political issues and viewed as a critical means of building trust and dampening security spirals. As Point Four of President Woodrow Wilson's famed Fourteen Points makes clear that peace required "Adequate guarantees given and taken that national armaments will be reduced to the lowest point consistent with domestic safety."[63] It seemed apparent to leaders at the time that the maintenance of large standing armies and major investments in military industries were associated with the militarism blamed for plunging the continent into war, and that a logical step toward lasting peace was the limitation of such activities. Thus, the disarmament of Germany and the defeated powers set strict quantitative measures on the manpower and materiel allowed to those states.[64] However, over the longer term, the Versailles Treaty and subsequent discussions and agreements envisioned a similar disarmament by the victorious states as well. This would become a consistent source of contention between Germany and its neighbors, ultimately providing the rationale from Germany's withdrawal from the Geneva Conference and eventually the League after Hitler's assumption of power.[65] Having been disarmed by the mandate of the victorious powers, subsequent regimes in Berlin during the period of the Weimar Republic viewed general disarmament as a necessary condition for any long-term, stable European diplomatic and political-military settlement. Leaders like Gustav Stresemann and Gerhard Bruning would repeatedly and forcefully advocate for "equality"—pushing the victorious powers to uphold their obligations and disarm "down" to Germany's levels. As the process unraveled in 1932–1933, calls for "equality" were used to introduce a rationale for allowing Germany to arm "up" to existing levels and be freed from League mandates if no cooperative solution could be found.

Yet while quantitative disarmament schemes could limit the numbers of men under arms and the configurations of those men (whether as professional militaries of mixes of professional and militia elements), and the materiel capabilities of a state, and would indeed be a component of any comprehensive resolution, the perceived impact of shifts in technology were already presenting problems for this approach.[66] The potential for dramatically increased mobility and firepower in the shape of tanks and mechanized vehicles, mobile artillery, and aircraft presented different kinds of challenges for advocates of disarmament that simple quantitative approaches could not adequately address. Leaders became increasingly aware of the destructive power of these new military applications and that they collectively threatened to shift the military balance away from the defense, making aggressive war possible and thus endangering the albeit uneasy status quo of post-Versailles Europe.[67]

A new approach emerged in the mid-1920s in response to these technological factors. This focused on a shared recognition of the post-World War I security environment as one that made offensive actions exceedingly difficult. As the carnage of the war exhibited, aggression would be difficult for any state without a fundamental change in the types of technologies used by the militaries of the day. The potential for such a change had been shown in the closing months of the war, as tanks, aerial bombing and increasingly powerful and mobile artillery offered the possibility of breaking through fortified defensive lines. However, if, as some political and military leaders argued, these new weapons could be abolished or limited to small numbers (making their deployment insignificant to the overall balance of forces), the defensive military environment could endure indefinitely. Further, if the defensive orientation of the system remained in tact, political disagreements could be settled without the fear of first strikes or fait accompli diplomacy.[68] Thus, diplomacy could be divorced from the threat of force, and conflicts could be negotiated and settled peacefully. In this way, these leaders echoed a line of defensive realism that would be articulated by scholars some 30 years later.[69] By focusing on the concept of *qualitative disarmament*, these leaders sought to incorporate legitimate state concerns about internal security and national self-defense within a framework that delineated between weapons that served defensive purposes and those that could be most efficacious for offensive or aggressive actions against neighbors.[70] While quantitative disarmament was not abandoned, the shift toward the definition of "aggressive" armaments became the centerpiece of disarmament discussions.

"capacity for aggressive employment" was the quality actually emphasized in the Geneva negotiations. . . . It assumes that war and insecurity result from the possibility of using military aggression as an instrument of national policy; hence it proposes to eliminate aggression by denying to national armed forces

> the weapons indispensable to the offensive. It envisages a future armament equilibrium in which actual resort to hostilities would end in a stalemate, and assumes that when the impossibility of successful offensive action becomes apparent, the tendency to attack will disappear.[71]

The potentially dramatic impact of new weapons made regulations concerned primarily with sizes and configurations of forces less important. What is perhaps more important is that the actors that took part in the Preparatory Commission and the conference itself had indeed been able to build consensus on the principle of qualitative disarmament. It was apparent to the diplomatic representatives and their military and technical advisors that certain classes of weapons were indeed likely to provide innovative states with the ability to unleash aggressive warfare against their neighbors. Limiting these weapons, and reinforcing the defensive nature of the postwar system would thus decrease the probability of war, since offensive operations would be perceived as prohibitively costly and thus aggression could be deterred without necessarily threatening neighbors. Leaders were deeply fearful of these weapons, and the military advisors saw their decisive value (and their potentially grave threat in the hands of others), therefore some system of effectively limiting their development and deployment would have to be devised, or else states could not forego acquiring them for their own security. Much of the Geneva Conference would focus on the development of such a system, which goes a long way in explaining its ultimate failure.

THE WORLD DISARMAMENT CONFERENCE AND ITS FAILURE

As representatives met beginning in the mid-1920s to develop a draft treaty on disarmament, a fundamental division between the participating states began to emerge that would, to a significant degree, structure the entire context of the subsequent discussions and proceedings. On one side stood Great Britain, along with the United States, and Germany. Their basic position was that disarmament, increasingly based upon the qualitative approach of prohibiting certain kinds of particularly offensive weapons, would provide the type of environment in which the larger issues of security could be discussed and successfully addressed.[72] Opposing this "disarmament first" camp was France and her allies of the "Little Entente" which consistently advocated a position that disarmament could only logically take place in the context of a stable diplomatic and political environment. That is, the underlying political issues that remained unresolved between France and Germany and the new states of Central Europe must first be concretely settled to the satisfaction of those parties before disarmament could be discussed.[73] The French position consistently clashed with that of Germany's. As the country that had lived up to its Versailles obligations of disarming down to a professional army of no more than 100,000 men and foregoing the development of new weapons, the Stresseman government expected the victors to follow its example as implied in the Versailles treaty.[74] Yet the French were adamant, throughout the period, to both empower the League of Nations, in terms of acquiring concrete guarantees of support from other members in the event of (ostensibly German or possibly Italian) aggression.[75] The division between the two camps carried into the conference as it began in February 1932.

FRANCE: THE QUESTION OF SECURITY OR DISARMAMENT

As the predominant continental power after World War I, it is not difficult to understand the French position.[76] Perhaps more importantly, memories of having been invaded by Germany twice in the past 50 years made attempts to assuage French fears of Germany difficult. Democratic regime or not, France was highly sensitive to any initiatives that would undermine its military superiority and thus its deterrent capability vis-à-vis Germany. Past experience and future uncertainty, not only in consideration of German intentions, but also over the availability of alliance partners and commitments of support against future aggression left France with a foreign policy in which a credible if not overwhelming deterrent capability was a central component.[77] Thus, France believed that it had a sincere interest in maintaining its own power, while perpetuating German weakness.[78]

> The French attitude toward arms reduction was entirely negative. France could not consider disarmament until it received concrete security guarantees.[79]

The inherent weakness in this underlying policy, however, was that it did not coincide with the stated policies of other great powers which soon viewed France as nearly as much of a "problem," nor did it recognize that, France was not interacting with a revisionist power, which would logically call for a deterrent strategy. While leaders of Weimar Germany may have questioned the equity of the Versailles Treaty, in fact during much of the 1920s and early 1930s, Germany could be considered a security-seeking state attempting to re-enter the system and primarily to address its own domestic difficulties.[80] Unfortunately, in adopting an overtly deterrent approach toward Germany, France had severe difficulties in interpreting signals in anything but a mistrustful, if not belligerent, light and undermined German leaders in the eyes of more revisionist domestic audiences. In this way, we can understand how even German pleas for equality leading up to and during the Geneva Conference could be interpreted as threatening, even though Germany only desired other states to follow through on their own disarmament obligations, and would accept further prohibitions to achieve that end.[81] Yet by adopting hardline policies, and refusing to consider German initiatives in good faith,

France itself sent signals to German leaders that were difficult to read in anything but a negative light.[82] This would be a constant source of frustration for Stresseman, Bruning, and their representatives, and it seems difficult to argue that this policy did not only provide fodder for the vehemently revisionist rhetoric of Hitler, but also exasperated and undermined the positions of moderate German elites.[83] More conciliatory policies toward Germany, similar to those taken by most of the other allies during the period may have helped to sure up the German domestic political situation and help the added benefit of a more unified allied line. Instead, the constant French intransigence not only frustrated Germany, but also its former partners.[84] The single most divisive political issue of the Geneva conference was that over "equality."[85]

The high level of distrust that drove French policy significantly colored its participation at the conference. Nonetheless, even against the ongoing French and German disagreement, it remains a fact that the political representatives of the various states were committed to agreement on disarmament, and qualitative disarmament seemed to provide a useful approach, even if it was only a starting point to address larger issues.

GREAT BRITAIN

Clearly, Great Britain was both publicly supportive of the objectives of the conference but also mistrustful of the process of League disarmament and French motives. Given its strategic position, it saw that the destruction of certain types of particularly aggressive weapons and the larger cause of disarmament could stabilize the continent, thus improving the probability that it would not have to once again enter into a European conflict. Facing a strong popular movement at home, the Labor government in power during the late 1920s saw disarmament as a worthwhile cause. Arthur Henderson was the the Labor Foreign Minister who was an initial driving force behind the Geneva Conference.

> The sheer horror of any future war made it inconceivable to him that a treaty would not be achieved. In his speech at the opening of the Conference he elaborated the familiar arguments for disarmament—the expense of huge military establishments, the insecurity caused by arms competition, and the rest. Almost passionately he pleaded: "The world wants disarmament. The world needs disarmament. We have it in our power to help fashion the pattern of future history. Behind all the technological complexities regarding man-power, gun-power, tonnage, categories, and the like, is the well-being of mankind, the future of our developing civilisation."[86]

Henderson lost his official position when the Labor government collapsed in late 1930. However, he continued to play an active role as the eventual president of the conference. Britain was a consistent supporter of qualitative disarmament throughout the period, supporting the prohibition of tanks and heavy caliber artillery, as well as bomber aircraft. Yet it was also cognizant of its own position as a maritime power, and was less willing to discuss naval prohibitions for fear of losing its (already declining) predominance over the seas.

A necessary political or diplomatic precondition for successful arms control cooperation may have indeed been a British guarantee of French security. Given the tenets of British policy, this was impossible. Despite the leadership role that British delegates took at the conference and the seeming willingness of the MacDonald Government to use arms control to improve British security by dampening naval tensions with the United States, the inability to cooperate with France would prove to be the central cause of the conference's failure.

> In practice, however, the government was unwilling to accept that the exertion of British power through direct participation in the European security system was an essential component in bringing about an agreement acceptable to both France and Germany. The Cabinet discussed the question on 15 December 1931, and again on 13 and 20 January 1932; but crucially it was decided that Britain could not accept commitment in addition to Locarno that might involve European war. The decision effectively sounded the death knell of the conference even before it had started.[87]

Without a concrete security guarantee from England, French leaders were unwilling to consider any proposal that would effectively allow Germany to achieve parity with France, whether such a plan entailed limitations of reductions of French forces and armaments or an allowance of German rearmament. British resistance to providing a guarantee only reinforced French advocacy for an agreement that would be dependent on empowering an international organization with the security of the continent.[88]

THE UNITED STATES AND OTHER POWERS

The official American position was similar to that of Britain's. While it too remained committed to the maintenance of a robust naval capability, it strongly supported the abolition of offensive land weapons. The impact of the 1929 Stock Market Crash had made spending cuts attractive, and given the strategic position of the United States, there was little need for large land forces or investment in newer military applications aside from their naval deterrent.

Perhaps the most vigorous and extensive case for the destruction of offensive weapons was advanced by the American delegation. The latter implied that such a step would serve two major purposes: first it would deprive a potential aggressor of the technical means required for successful attack under modern conditions, thus reducing the profits of aggression; second, it would by rendering sudden attack and defeat obviously impossible, allay the psychological condition of "insecurity" underlying disturbed political relations and governing the progressive increases in armed forces. This argument tacitly assumed that while to some extent fear causes armament, to a much greater extent armament causes fear.[89]

It is important to note that the other major powers adopted constructive approaches, regardless of their regimes or ideologies. The Soviet Union had through much of the 1920s called for complete and general disarmament, and the initial speech by Stalin's representative to the conference, Maxim Litvinoff echoed this policy.[90] However, given the general distrust of the Bolshevik regime, these calls for complete disarmament were seen as disingenuous and merely another example of communist propaganda. The persistent fear of communist subversion, embodied by the existence of the Comintern, interpreted these calls as an all-too-convenient request for European states to degrade their capabilities to maintain domestic order. Yet after Litvinoff's initial calls were expectedly rebuffed, the Commissar for Foreign Affairs quickly moved to constructively engage his fellow representatives in seemingly honest attempts to develop concrete measures of qualitative disarmament, generally supporting recommendations of the British camp. As the conference floundered, he began to become more receptive to the French position, setting the groundwork for what would become his (eventually futile) attempts to build collective security measures against Hitler.[91] Like his Soviet counterpart, Foreign Minister Grandi of Italy consistently supported qualitative disarmament and the proposals of Britain and the United States in particular. Though the Mussolini's subsequent revisionism calls into question the sincerity of the Italian government, at the time, Rome and its representatives were considered to be cooperative partners in the attempt to reach agreement on the limitation of aggressive armaments.[92]

THE TARDIEU PLAN: A CALL FOR "INTERNATIONALIZATION"

After consideration of the draft treaty presented by the Preparatory Commission, which was little more than a general framework encompassing the questions to be addressed by the conference, France focused its concern on guarantees that prohibited weapons would indeed be off limits. More importantly, France demanded that in the event of renewed aggression, states who had abided by an eventual treaty would not be left unprotected. The French response to the draft treaty came in the form of a proposal offered by the French Foreign Minister, Pierre Tardieu on February 5, 1932. This plan called for the placement of civil aviation, long-range artillery, and heavy warships under League control. The plan also envisioned the development of an international police force under the League command and "the acceptance of compulsory international arbitration in all disputes, with the term 'aggression' specifically defined."[93] In effect, the French were more realistically acknowledging the inherent difficulties of controlling new technologies, admitting that the further development of civil aviation under the control of individual states would only perpetuate the possibility of exploiting that capability for military means. Moreover, the inclusion of the enhancement of League capabilities and dispute resolution procedures reiterated the underlying French commitment to tie participating states to clear commitments of collective security and standards of behavior.[94] The French believed that "armaments could be secretly manufactured and stored in significant quantities" and that "there was no sharp dividing line between war materials and peace materials" and "if a world force had charge over the more powerful arms, it would be able to overcome armies fighting with only the converted machinery to use as weapons."[95] Abolition would not guarantee security because states bent on aggression could engage in clandestine development and innovation, while states that abided by an abolition agreement would be at their mercy.

As a continental state with an unfortunate recent history being the victim of aggression, France was keenly sensitive to the possibility of technology providing new opportunities for aggression even in the aftermath of a successful disarmament treaty. In delivering his plan, Tardieu underscored the competitive nature of military innovation and saw little value in mere prohibitions or limitations—whether qualitative or quantitative—without a robust oversight mechanism and some political agreement on security guarantees.

> We shall find that the engineers who, unfortunately, are not our allies in our work of limiting armaments, but who are always striving to obtain more powerful means of destruction for a smaller expenditure of money, will cause us some unpleasant surprises. We have all heard mention—and I fully appreciate the skill of German engineering—of what is called the pocket battleship. Do you suppose that the engineers will forbid themselves to devise after the pocket battleship, the pocket siege-gun or the pocket aeroplane? Of course they will not.[96]

Tardieu's concerns clearly illustrate the impact of dynamic technology upon existing military environments. Even during a period of clear superiority, in relative power terms, France could not commit to any agreement that would not guarantee her security precisely because she feared that the dynamic nature of technology would provide opportunities for potential enemies in the future. Even with agreement, including the possibility of verification under a multilateral authority, the perception of rapidly changing technology during the period presented insurmountable difficulties. Moreover, the remedy, the notion of "internationalization" was quickly condemned by most of the participants for obvious reasons. Neither Great Britain, nor the United States, nor Japan was willing to see their naval assets taken out of their control and given to any other authority or degraded, particularly given the highly uncertain security environment in the Pacific. Germany simply viewed the proposal as another attempt by the French to avoid the question of equality, adding to their frustrations. Under the French plan, Germany would remain in a position of weakness.

THE BRITISH DEBATE

As mentioned above, within the British government, significant divisions emerged over the question of disarmament. While support for qualitative disarmament was the official government line, many influential British political and military leaders disagreed with the concept and remained skeptical of its utility. Most notably, Sir Winston Churchill, who would serve as Chancellor of the Exchequer during the period and William J. F. C. Fuller a leading military intellectual who foresaw the use of armor in the war to come, both argued that weapons could not be distinguished and any such attempts were futile. New weapons could be used equally well for offense or defense, and limiting certain categories would only lead to development of different applications for new technology that presented similar capabilities. Moreover, given Britain's overseas commitments, it seemed imprudent to prohibit armaments that could be potentially useful in maintaining its power and influence.[97]

Opposing this view was Henderson, Lord Cecil, the British representative, and military thinker Basil Liddell-Hart. Liddell-Hart, who like his mentor Fuller, understood better than most the capabilities offered by armored warfare, and also that these weapons could be utilized defensively, nonetheless focused on their ability to initiate war. Moving away from the question of the utility of these weapons in an actual conflict, Liddell-Hart focused instead upon the idea that certain weapons could be effectively used to *launch* aggression. While tanks and heavy artillery could certainly be used for defensive operations, they could also be used (particularly with bombers) to launch rapid and decisive strikes against a defense on the first day of war.[98] If the rationale behind the conference was to develop an international environment that dampened the potential for future wars by focusing on the abolition of certain weapons, then starting with those most capable of contributing to decisive first strikes and thus undermining the stability of the system, made perfect sense. During the period Liddell-Hart wrote:

> And I cannot see how anyone can deny that the abolition of heavy artillery and heavy tanks would not be a definite handicap to the offensive, and so a discouragement to aggression.[99]

A recollection of Liddell-Hart's experience as a technical advisor for the Preparatory Commission and the conference captures the debate within British defense policy circles and is worth quoting at length:

> Liddell Hart was one of the few military experts who took the logical step from propagating the tank and bomber as means of enhancing the offense and restoring decisiveness to war, to arguing for their abolition as a means of war prevention. . . . "If I propounded the 'disarmament antidote' to it [his former support for the development of armored warfare] and helped to obtain its adoption at the coming conference, it would mean strangling my own 'baby.' Others were not prepared to commit infanticide. Thus Fuller and Winston Churchill opposed Liddell Hart and the proposed abolition of tanks. . . . [Liddell Hart] gave Fuller credit for developing a method 'which rescued soldiers from the abysmal prospect of further Sommes and Paschendaeles.' This he agreed was the general's task. . . . The statesman's aim, Liddell Hart argued, was that of preventing war, rather than winning it. And this aim required not the strengthening of offense to produce a quick victory, but exactly the opposite, that of enhancing the defense.[100]

This division reflects a larger dynamic at work in both the preparation and execution of the conference. While the political representatives consistently agreed with the theme of qualitative disarmament, the technical commissions and subcommittees ostensibly charged with negotiating the details and providing expert advice found no such common ground.[101] Echoing the positions of Fuller and Churchill, the military experts that made up these commissions consistently fought for reservations and exceptions, making consensus, never mind unanimity, impossible.

After a prolonged debate on whether a distinction should be made between defensive and offensive weapons in discussing the problem of disarmament, they decided that "while it would be most useful from the practical point of view to pose the question, it is almost impossible from the military technological point of view to distinguish between weapons in terms of dividing them into offensive and defensive ones, for with the present-day development level of technology, nearly every weapon (with the possible exception of immobile fortress installations, gun sites, and so on) can be used in an offensive war.[102]

In what could be viewed as self-defeating or short-sighted, those most sensitive to the dynamic nature of technology and its implications for national security simply were not willing to agree to commit their states to a treaty or regime that could *potentially* leave them worse off vis-à-vis potential enemies in the future. A successful resolution of the disarmament question may have indeed decreased the probability of a future conflict, but nonetheless defection was seen as preferred strategy as a hedge against a possible future adversary which could be expected to exploit these new technologies.

Perhaps more troubling, it became increasingly clear to observers Britain was not alone in having this crucial debate and that that "those in each Power responsible for foreign and defense policy believed that they could not surrender any advantages they might have in possessing armed forces unless guarantees of security existed."[103] Because of this, and the political disagreements concerning German equality at the plenary sessions, the conference adjourned in March 1932 with little agreement aside from a general resolution supporting the principle of qualitative disarmament.[104] Thus when the conference recessed for Easter in March 1932, there were certainly still contentious issues to be discussed and resolved. But one thing that had become apparent was that the concept of "aggressive" armament had been "simply and easily" defined in general terms and had been accepted by a large consensus of the participants.

> An analysis of the different proposals and statements shows that 14 countries supported the abolition of heavy long-range artillery; 11 favored the abolition of tanks; 13 the abolition of bombing aircraft; 7 the abolition of all military aircraft; 7 the abolition of all capital ships; 6 the abolition of aircraft carriers.[105]

Various states continued to press for certain exceptions or allowances under more comprehensive disarmament plans, but it was clear that the general contours of a qualitative disarmament scheme had been developed through the course of negotiations.[106] The head American delegate, Hugh Gibson focused his energies on gaining support for an American proposal that abolish "weapons of a nature peculiarly offensive against land defenses."[107]

> "I am convinced," said he, "that in abolition of these weapons, lies the key to the whole difficulty in the war of disarmament, that of removing fear and providing security."[108]

The specifics of the plan focused explicitly on "Tanks, heavy mobile guns, and gases" which should be abolished.[109] While Gibson's plan received support from the delegates of Great Britain, Germany, and Italy, France's Tardieu opposed the specific focus on land armaments, and more generally the notion of abolition. Aircraft and naval vessels must also be included in any comprehensive agreement and the requisite supervision and oversight as well as sanctions for potential aggressors must be addressed. The former point received support, including from the United States, while the latter—the call for internationalization and the placement of armament under the supervision of a League entity—would serve as the fault line throughout the conference.[110]

TOWARD THE "APRIL TRAGEDY"

Fearing for his position in the wake of Nazi gains in provincial elections, Chancellor Bruning made an offer directly to British Prime Minister Ramsay MacDonald and U.S. Secretary of State Henry Stimson who had agreed that the negotiations had lost momentum and that direct intervention at Geneva may facilitate an eventual agreement.[111] French Prime Minister Tardieu was in the midst of campaigning prior to the May general election but was expected to return to Geneva, particularly if a breakthrough was reached. Bruning's proposal made significant concessions, including a five-year moratorium on any increase in German armaments (or until a subsequent conference), an agreement to prohibit offensive weapons should the larger conference agree to similar prohibitions, as well as reductions in service time and the organization of a militia along the line set out in the Treaty for the Reichswehr.[112] The proposal was well received by MacDonald and Stimson, as well as Italy's Minister Grandi. Tardieu's return to the Geneva was requested, but he subsequently failed to attend. It remains somewhat of mystery to this day why Tardieu did not return given the stakes involved and the foreseeable consequences. With the subsequent publication of Bruning's last attempt to achieve equality and its failure, he was removed from office by President Hindenburg on May 30 after his return to Germany and was replaced by von Papen and what has become known as the "April Tragedy" was complete. The French elections also removed Prime Minister Tardieu from the disarmament discussions.[113]

Nonetheless, even in the face of this considerable hindrance, the political representatives continued to struggle to develop solutions that could both address the objectives of disarmament and the political difficulties of integrating within a larger (primarily French) notion of security.

THE HOOVER PLAN

After the inconclusive and ineffective end of the first stage of the conference, President Hoover attempted to seize the initiative and presented a sweeping plan calling for dramatic disarmament based upon qualitative principles.[114] However, though it was well-received by both world public opinion and advocates of disarmament, it was far less enthusiastically received by the representatives of the other great powers. "Thus, because Hoover's suggestions were made without any consideration for the strategic concerns of the other Powers, they were politely rebuffed."[115] As it would happen, President Hoover's plan was one of the last, best chances to save qualitative disarmament and salvage the conference. At the same time, the policy of the United States was not beyond charges of hypocrisy.[116] However, this seeming hypocrisy reflects a larger dynamic that hindered the work of the conference. Different states with different security objectives have needs for differing types of military capabilities thus making consensus difficult. "But as important as organizational matters were, the reality of arranging general disarmament derived from reconciling the political and strategic interests of the Powers: their need to equate different types and numbers of weapons and, then, meet their unique military requirements concerning the political both international security and national sovereignty."[117] New decisive military applications offered states specific opportunities to improve their security and thus in an uncertain environment, were difficult to forego.

THE CONFERENCE APPROACHES A CONCLUSION

The first two stages of the conference failed to reach agreement, save several general resolutions and various speeches supporting general principles of qualitative disarmament. In reality, nothing concrete was accomplished, as political leaders struggled to find agreement on the difficult questions of verification and definitions and attributes of armaments to be prohibited. "In fact, to a large extent, the conversations at every level of the World Conference were going over ground covered ad nauseum in the six years of the Preparatory Commission."[118] One general conclusion was that abolition, rather than internationalization was the only way to move forward. This did not reflect the fact that abolition was viewed as effective, only that internationalization was politically impossible.[119] At the time President Hoover made his bold attempt to salvage the conference, French Minister Herriot offered another French proposal that, for the most part, simply rehashed earlier positions.[120] The essential French distrust of any policy leaving abolition up to the good will of other states drove their subsequent policies even after internationalization was rejected. It advocated for the development of a permanent international authority, with significant powers, to monitor and verify the armament activities of all countries, and a double standard that would apply different regimes to the victorious and defeated powers. Obviously, this latter proposal only exasperated the German representatives and effectively dashed further discussion.[121]

While the British expressed muted public support for the Herriot Plan, ultimately the question of security guarantees precluded a full endorsement and more time was spent discussing qualitative reductions of forces and requisite surveillance.[122] Attempting to salvage the conference, or at least Britain's image as a leader and support its objectives, Prime Minister MacDonald unveiled a plan on March 16, 1933, that would have allowed for German equality, provided a "security and consultative pact" to address any future conflict or crisis and a convention that would have prohibited artillery, tanks, aircrafts, and naval vessels. However, it lacked an all-important guarantee of French security.[123]

> Under the terms of the MacDonald Plan, France would be required to reduce her military superiority over Germany, and it was scarcely conceivable that she would agree to this without increased commitment from Britain, at the very least. Unfortunately, the British Cabinet was becoming almost as irritated with the ever increasing French demands for security as they were with Germany's changing definition of equality.[124]

On October 14, 1933, Chancellor Adolf Hitler officially withdrew from the Conference, reiterating German commitments to peace and disarmament and to the cause of achieving equality.[125]

Ultimately, the conflict between abolition and internationalization reflects the underlying difficulty of disarmament and strongly supports MET. In an anarchic system of sovereign states, one can never be certain of the motives or activities of one's neighbors. The French certainly felt this way, with some legitimate reasons. The need for verification and the general concern that technology would provide states with opportunities to increase their capabilities to inflict decisive attacks on their neighbors seems essential for the success of any meaningful form of arms control. Moreover, the difficulties presented by the increasingly prevalent civilian applications of new technologies (particularly planes and motorized vehicles) made the idea of abolishing their military applications all the more difficult to accept. To dispel the uncertainty created by the dynamic nature of technology, some form of surveillance, verification, and (ideally) sanctioning components would seem to be necessary preconditions before any state could agree to let its own guard down.

In an anarchic environment, states must provide first and foremost for their own security. "Apart from the technical considerations of balancing air, land, and sea weapons, there was a political dimension intimately connected with military policy and planning. . . . Ultimately, compromise proved to be a mirage. No one was willing to jeopardize the strength of their national armories for political trade-offs."[126] Without guarantees that future technologies would not adversely affect their security, states could not agree to limit their own access to technologies that could prove critical to winning a coming conflict. Thus, in the absence of a hegemonic or supranational power with real capabilities (which the League was certainly not), states were not likely to agree to disarm.

At the conclusion of the second stage of the conference, the political environment took a significant shift for the worse. The eventual ascension of a Nazi government was a serious blow. Though Hitler did not immediately pull Germany out of the conference, preferring to maintain a cooperative posture in the short term, it was clear the time for agreement had passed. At the same time the Duce was shifting his own posture toward the League and the conference, and Grandi, who was generally viewed as a constructive delegate, was replaced.[127] While the plenary sessions continued to serve as a venue for the discussion of disarmament, there was little hope for an agreement. The security situation in Europe only became more tenuous after Hitler's rise, despite his initially appeasing signals. For all intents and purposes, the conference was over with the fall of Bruning.

ASSESSING THE RELATIVE INFLUENCE OF TECHNOLOGICAL AND POLITICAL FORCES

As the discussion of the Conference indicates, it is difficult to delineate the role of conflicting state interests from the role of technological uncertainty in attributing ultimate causality for its failure. The French obstinacy, which in truly self-defeating ways, served to isolate it from its former allies and provide the charges of double standards already emerging from German and Italian revisionists with some level of credibility, if not legitimacy, is difficult to overestimate.[128] However, the real question is this: Had the French been more constructive and aligned its policies with Great Britain and the United States, and thus moved toward a real rapprochement with Germany, would the conference have succeeded? It seems unlikely. It certainly may have been more positive in tone, and resulted in some general agreements on the underlying desirability of qualitative arms control, however the fact remains that coming to consensus, never mind concrete agreement, on the limitation of armaments is exceedingly difficult. Given the uncertainty of the political and military environments, and given the highly uncertain potential benefits and challenges presented by the utilization of emerging technologies into new military applications, a real comprehensive settlement was unlikely.

Given the history of Europe, the dynamics of an ongoing security dilemma are apparent.[129] Reassurance may require quite long periods of time, involving numerous and repeated interactions, and though the Weimar governments under Stresseman and Bruning consistently signaled a willingness to cooperate in disarmament and the overall mission of the League, lingering mistrust over Germany's past and fear of the future made France unwilling or unable to interpret these signals and reorient its policies appropriately. This does not make French policy irrational, even as it seems self-defeating in hindsight. It merely reflects the difficulty of cooperation is an environment of mistrust and uncertainty. The fact that both the United States and Britain desired a stable Europe, and one that could survive against the continuing internal threats of Bolshevism, but that neither were willing or able to translate those desires into a real commitment to the continent exacerbated France's real sense of insecurity. Likewise, the Soviet Union, who almost by definition saw itself as besieged by unfriendly if not hostile capitalist states, faced a daunting task to become a "normal" state, despite the best attempts by Maxim Litvinoff. This would become clear later when attempts to build a collective security arrangement against the rising Nazi threat were consistently rebuffed by Britain and France, not to mention its smaller neighbors. Against this political backdrop, disarmament was difficult.

Despite the recriminations of observers at the time of the conference's conclusion or since, the discussion of this chapter clearly shows that the respective representatives and leaders did in fact achieve a relatively clear picture of what was likely to be "offensive" and most likely to overturn the defensive security environment that emerged from the Great War.[130] What France, Britain, and the other states rejected was an agreement that would foreclose their ability to develop and deploy new and decisive offensive weapons precisely because of their expected benefits and the grave threat those weapons would present if deployed by a future enemy. Even with wholly defensive motives, states will continually search for ways to better address those needs and new technologies hold out the potential of doing so in more effective ways. At the same time, new technologies can be perceived as threatening to others, or provide capabilities that, in terms of particularly "offensive" capabilities, do in fact present real, objective threats to the security of others. Thus, the asymmetry between the types of weapons desired by different states provided an increased difficulty for general disarmament.[131] Beyond this, distinguishing between the offensive and defensive nature of technologies can be difficult and particularly in the presence of underlying security dilemmas or conflicts of interest between states which these technological problems can exacerbate.[132]

If it were possible to "remove" technological change—and the emergence of decisive military applications—from our analysis, thus leaving the defense-dominant security environment of World War I intact, where aggression was universally understood to be prohibitively costly and where states could provide for their security without threatening their neighbors, the history of the period could have been quite different. Obviously, this counterfactual is difficult to envision, but if the means to launch and wage a decisive first-strike against one's neighbors was not available or within the realm of possibility, it seems likely that the intense insecurity faced by states would have been considerably dampened. This is not to say that war was impossible. Instead, it simply shows that the dynamic nature of technology and its potential to provide states with new military capabilities can have a significantly destabilizing effect on even the most stable of defense-dominant systems. Leaders must be aware that in the future they could face a very different security environment because of changes in technology.[133]

CONCLUSIONS

The ultimate failure of the Geneva Conference on World Disarmament can be explained by realist balance of power concepts, most importantly, the failure to adequately address the underlying disputes created by the Versailles Treaty. But this broad characterization seems to overlook a number of important ideas that not only tell a more complete and intricate story of failed diplomacy, but also have significant implications for the study of international relations theory more broadly. As this chapter illustrates, and MET would predict, the dynamic nature of technology and its potential translation into decisive new military applications undermined the ability of military and political representatives to come to consensus on what (and how) to abolish in order to maintain a "defense-dominant" system where aggression would be prohibitively costly.

The puzzle presented by the World Disarmament Conference for defensive realism is not why it failed in 1933, but why it failed so miserably in the previous eight years to make any headway toward serious limitations. By the time Hitler came to power in Germany, there was little hope of any meaningful multilateral cooperation. However, the Preparatory Commission of the World Disarmament Conference first convened in 1928. Certainly during the mid-1920s, political conditions in Europe would have seemed favorable for arms control cooperation. The discussions and debates that took place within the technical committees and political plenary sessions provided the term "qualitative disarmament." As defensive realism would predict, the delegates to the conference worked to effectively ban or control offensive weapons, or weapons that could be used to execute offensive operations, while allowing states to maintain defensive weapons. In doing so, they were essentially attempting to lock in the defense dominant world that had seemingly been left in the wake of the First World War. Moreover, contrary to the notion that weapons cannot be categorized as simply offensive or defensive, the technical experts and sympathetic delegates did just that, focusing on the applications that would most likely be used to return maneuver to modern warfare. Nonetheless, no compromise could be forged.

The challenge was not in devising categories of offensive versus defensive weapons. The critical issue the conference bogged down over was the means by which offensive weapons would be addressed under an arms limitation regime. Two alternatives were offered and discussed. Great Britain argued for abolition of those weapons deemed to have offensive potential. Conversely, France argued that all such offensive weapons should be placed under the control of the League of Nations. Heavy tanks, military aircraft, mobile artillery should be "internationalized" and placed under the control of League monitors. Ostensibly this would serve to remove these from national arsenals, decreasing the probability of conflict or war, while also strengthening the League's capabilities and therefore improving its deterrent against future aggression. Of course, outside of France, such an idea had little support.

A necessary political or diplomatic precondition for successful arms control cooperation may have indeed been a British guarantee of French security. Given the tenets of British policy, this was impossible. Despite the leadership role that British delegates took at the conference and the seeming willingness of the Baldwin Government to use arms control to improve British security by dampening naval tensions with the a United States, the inability to cooperate with France would prove to be the central cause of the conference's failure. With the benefit of hindsight, the policy seems remarkably short-sighted. Great Britain effectively allowed the conference to collapse, and while it did not possess the resources to exploit the new military applications of technological change (Britain abandoned the Experimental tank force after extensive forward looking exercises that would heavily influence German ideas on armored warfare and mechanization) it would soon face the desperate challenge of responding to Germany's rearmament in the late 1930s.[134]

Both the British and French positions exemplify the logic of MET. Where cooperation (before Bruning's fall and Hitler's assumption of power) may have significantly improved the security of both Great Britain and France, and also reassured a hesitant and troubled Germany, neither could agree to forgo decisive weapons. Even if the status quo powers were unlikely to focus resources on maximizing the potential capabilities of these new weapons (as the troubled powers would) they were simply too frightening to forgo in an uncertain security environment.

However, as discussed above, the failure of the World Disarmament Conference at Geneva is over-determined. An offensive realist account would seemingly offer a general account of the inability of great powers to cooperate during a time of technological innovation with significant perceived military implications, as well as a deteriorating security environment. But such an account would be far too simplistic and ignore the actual history of the Conference and the work undertaken by the Preparatory Commission. For MET, the experience provides strong support for its first general proposition which would predict failure in the presence of decisive weapons, also explains the motivations of leading states to broach the possibility of qualitative arms control under relatively favorable political-military conditions. However, because of inherent difficulties in confidently monitoring any limitations of reductions, cooperation will fail. It also supports the specific hypothesis that considers the relative simultaneity of the developments and deployments of decisive weapons. The general acceptance of plans, whether for abolition or internationalization based on the principle of qualitative disarmament, seriously questions the relevance of defensive realism in this case. The nature of the environment under which cooperation was initiated would seemingly have provided a real opportunity to "lock in" a defense-dominant world by proscribing armaments that could be used for offensive operations. However, as MET would argue, this represents a static depiction of the world and the role of technological change clearly undermines the applicability of defensive realism here. Given significant role of a the League of Nations in this case, as well as the technical and informational support given to leaders by various League entities, the failure of the conference severely undermines the contribution of institutional approaches to our understanding of security cooperation. Finally, the cooperative predictions derived from the presence of significant domestic resource constraints or shared values, ideas or worldviews would seem to perform poorly in this case. The decisive potential of new weapons drove states to avoid foreclosing their eventual development and deployment, even if they had significant challenges in acquiring them. Moreover, despite their shared values and interests in maintaining the status quo in Europe, the differences in policies employed by the great powers ultimately proved far more important to the story of the Geneva Conference.

NOTES

1. Technological advancement together with the capabilities of states to mass-produce new weapons fundamentally shifted the nature of warfare. For a discussion of the various military development programs that took place between the wars, see Williamson Murray and Allan R. Millett, *Military Innovation in the Interwar Period* (Cambridge: Cambridge University Press, 1996); and Williamson Murray and Barry Watts, *Military Innovation in Peacetime* (Paper Prepared for Director of Net Assessment, Mr. A.W. Marshall, 1995). Also see Michael Geyer, "German Strategy in the Age of Machine Warfare, 1914–1945," 527–97; Brian Bond and Martin Alexander, "Liddell Hart and De Gaulle: The Doctrines of Limited Liability and Mobile Defense," 598–623; and David MacIsaac, "Voices from the Central Blue: The Air Power Theorists," 624–47, all in Peter Paret (ed.), *Makers of Modern Strategy from Machiavelli to the Nuclear Age* (Princeton, NJ: NJ: Princeton University Press, 1986).

2. Concepts of strategic bombing, which would target the enemy's homeland and infrastructure was seen as providing a way to win quickly and decisively. At the same time, this was viewed as more humane than the horrific losses suffered in the trenches of World War I. See, for example, Ward Thomas, *The Ethics of Destruction, Norms and Force in International Relations* (Ithaca, NY: Cornell University Press, 2001), chap. 4.

3. On the Geneva Conference and the history of disarmament initiatives throughout the 1920s, see the following contemporary coverage: "The Disarmament Deadlock" in *Foreign Policy Reports* 4:19 (New York: Foreign Policy Association, 1928), 381–96; Mildred S. Wertheimer, "The League of Nations and the Prevention of War," *Foreign Policy Reports* 6:11 (New York: Foreign Policy Association, 1930), 206–24; David Woodward, "Limitation of Land Armaments," *Foreign Policy Reports* 6:2 (New York: Foreign Policy Association, 1930),

17–35; David Woodward, "Limitation of Air Armaments," *Foreign Policy Reports* 6:17 (New York: Foreign Policy Association, 1930), 295–313; William T. Stone, "The Draft Treaty for the World Disarmament Conference," *Foreign Policy Reports* 6:25 (New York: Foreign Policy Association, 1931), 469–88; William T. Stone, "The World Disarmament Conference, First Stage, February 2–March 17, 1932," *Foreign Policy Reports* 8:5 (New York: Foreign Policy Association, 1932), 60–66; and William T. Stone, "The World Disarmament Conference, Second Stage, March 17, 1932–January, 1933," *Foreign Policy Reports* 8:23 (New York: Foreign Policy Association, 1931), 267–78. For analysis of the disarmament movement and the conference, see Denys P. Myers, *World Disarmament: Its Problems and Prospects* (Boston: World Peace Foundation, 1932); John W. Wheeler-Bennett, *Disarmament and Security Since Locarno, 1925–1931* (London: George Allen and Unwin, 1932); Wheeler-Bennett, *The Pipe Dream of Peace, The Story of the Collapse of Disarmament* (New York: William Morrow, 1935); and Philip Noel-Baker, *The First World Disarmament Conference 1932–33 And Why It Failed* (Oxford: Pergamon Press, 1979).

4. Many of the military experts who served on the technical committees that ostensibly advised the political representatives were also leading military thinkers of the time and foresaw the military benefits and challenges presented by the emerging military applications of new technologies in a future war.

5. The resolution passed by the conference, the so-called Benes resolution, which effectively adopted the concept of qualitative disarmament reflected this consensus. Unfortunately, the translation of this general principle into concrete prohibitions failed. See William T. Stone, "The World Disarmament Conference, First Stage, February 2–March 17, 1932."

6. The offense-defense balance has been a central component in the theory of defensive realism. Important work on this topic includes: Robert Jervis "Cooperation under the Security Dilemma," *World Politics 30*, no 2 (1978); Charles L. Glaser and Chaim Kaufmann, "What Is the Offense-Defense Balance and Can We Measure It?" *International Security 22*, no. 4 (Spring 1998); Stephen Van Evera, "Offense, Defense and the Causes of War," *International Security 22*, no. 4 (Spring 1998); and Sean Lynn-Jones, "Offense-Defense Theory and Its Critics," *Security Studies 4*, no. 4 (Summer 1995). Jack S. Levy questions the utility of the offense-defense balance as a conceptual tool in "The Offensive/Defensive Balance of Military Technology: A Theoretical and Historical Analysis," *International Studies Quarterly 28* (1984): 219–38. For an interesting debate on the subject see "Correspondence: Taking Offense at Offense-Defense Theory," *International Security 23*, no. 3 (Winter 1998/99).

7. Spencer C. Tucker, *The Great War 1914–1918* (London: University College Press, 1998), 16.

8. See, for example, J. F. C. Fuller, *The Conduct of War 1789–1961* (New Brunswick: Da Capo Press, 1962), 151–82; William H. McNeill, *The Pursuit of Power* (Chicago: University of Chicago Press, 1982), 262–63.

9. There was greater maneuver on the Eastern front due to the open terrain. See Fuller, *The Conduct of War*, 167.

10. On the "Cult of the Offensive," see Stephen Van Evera, "The Cult of the Offensive and the Origins of the First World War," *International Security* 9, no. 1 (Summer 1984); Jack Snyder, "Civil-Military Relations and the Cult of the Offensive, 1914 and 1984" *International Security* 9, no. 1 (Summer 1984). Also see Scott D. Sagan, "1914 Revisted: Allies Offense and Instability," *International Security* 11, no. 2 (Fall 1986).

11. On the Schlieffen plan and its execution, see Tucker, 22. "German historian Gerhard Ritter has noted that the Schieffen Plan, while bold, failed to take into account the French rail net. German troops would have to move on foot along their great arc. Schlieffen's plan allowed no deviations in timetable, yet the invaders would inevitably be slowed by blown bridges and lengthened supply lines. French defenders inside the arc, meanwhile, would be able to utilize railroads to shift resources quickly along lateral lines. The Schlieffen Plan might have worked either earlier or later: in Napoleonic times, before the advent of the railroad; or later during the Second World War, when mechanized vehicles could rapidly transport large numbers of men and supplies across open country and aircraft could disrupt enemy mobilizations and troop movements."

12. See Gunther E. Rothenberg, "Moltke, Schliefen, and the Doctrine of Strategic Envelopment," in Peter Paret, ed., *Makers of Modern Strategy from Machiavelli to the Nuclear Age* (Princeton, NJ: Princeton University Press, 1986), 296–325.

13. See Michael Howard, *The First World War* (Oxford: Oxford University Press, 2002). The so-called Nivelle's Offensive of April 16, led directly to 130,000 French casualties and a collapse into "civil strikes," see pp. 99–100. The Ludendorff Offensive, March 1918, was similarly devastating. By the end of April it had led to the death of 350,000 men, and another 130,000 by May. See Howard, pp. 119–23.

14. See Keir A. Lieber, "Grasping the Technological Peace," *International Security 25*, no. 1 (Summer, 2000), 71–94; and Jonathan Shimshoni makes this point in "Technology, Military Advantage and World War I, A Case for Military Entrepreneurship," *International Security 15*, no. 3 (Winter 1990/91).

15. Jonathan B. A. Bailey, "The First World War and the Birth of Modern Warfare," in Knox and Murray, *The Dynamics of Military Revolution 1300–2050*, 144–45.

16. Tanks were predominantly utilized in the role of infantry support. This was successful in the closing battles of the war, as the British used combined armored-infantry assaults to break the German lines in August 1918, but that success also reflect the degradation of German forces at that point. In fact, tanks were also successfully used defensively to plug holes in allied lines during the previous spring. For a discussion, see Tim Travers, "Could the Tanks of 1918 Have Been War-Winners for the British Expeditionary Force?" *Journal of Contemporary History* 27, no. 3 (July 1992): 389–406.

17. Marion William Boggs, *Attempts to Define and Limit "Aggressive" Armament in Diplomacy and Strategy* (Columbia: University of Missouri Press, 1941), 76–77.

18. This is a difficult counterfactual argument to make given the exhaustion of all sides by 1918. However, had tanks been introduced in significant numbers without a development of anti-tank weaponry, it is conceivable that the battlefield may indeed have become one of maneuver, or offense-dominance. See Travers, "Could the Tanks of 1918 Have Been War-Winners for the British Expeditionary Force?" Similarly, the use of artillery has been credited with offering the potential to shift the nature of the war: Jonathan B. A. Bailey, "The First World War and the Birth of Modern Warfare," in MacGregor Knox and Williamson Murray, eds., *The Dynamics of Military Revolution 1300–2050* (Cambridge: Cambridge University Press, 2001).

19. It is difficult to find in recent history a world that was more "doubly safe" as that following the First World War. States were able to both recognize the defense dominance of trenches, fortifications, machine guns and barbed wire, and were also able to differentiate these as "defensive." To overturn this system would require the development, acquisition, and deployment of new weapons systems capable of overturning this balance. Jervis discusses the notion of a "doubly safe" world in "Cooperation under the Security Dilemma."

20. Writing about the Great War clearly illustrates the universal understanding of the horrors of the war and the tremendous human costs. Moreover the sense of futility that accompanied the deaths because of the positional nature of the war only reinforced the notion that it should be the "war to end all wars." See Howard, *The First World War*, and his piece "Men against Fire: The Doctrine of the Offensive in 1914," in Paret, ed., *Makers of Modern Strategy from Machiavelli to the Nuclear Age*, 510–26.

21. Jervis first made this argument in "Cooperation under the Security Dilemma," *World Politics 30*, no 2 (1978), and other scholars in the traditions of defensive realism have followed suit. See, for example, Jeffrey W. Taliaferro, "Security Seeking under Anarchy," *International Security 25*, no. 3 (Winter 2000/01); Charles Glaser, "Realists as Optimists: Cooperation as Self-Help," *International Security 19*, no. 3 (Winter 1994/95); and Stephen Van Evera, *Causes of War, Power and the Roots of Conflict* (Ithaca, NY: Cornell University Press, 1999). As Mearsheimer correctly implies, though, technology is not static, and new technologies will provide states with new opportunities that could overturn a defense dominant world. It is hard to argue that the European system after World War I was not defense dominant. The combination of trenches, machine guns, and barbed wire made large-scale offensive operations prohibitively costly. Yet attempts throughout the 1920s and early 1930s to "lock-in" this defense dominant world failed, setting the stage for thestate to engage in the development and deploy-

ment of new technologies (such as armor and strategic bombing) and strategies (combined arms warfare) that made offensive operations possible and effective. In this way we can begin to see technology as a dynamic variable that holds out the possibility of eroding or even overturning a given offense-defense balance at a given time. For an interesting critique of defensive realism, see Keir A. Lieber, "Grasping the Technological Peace," *International Security* 25, no. 1 (Summer 2000).

22. See, for example, Paul Kennedy, *The Rise and Fall of the Great Powers: Economic Change and Military Conflict from 1500 to 2000* (New York: Random House, 1987), 263–88.

23. Though the entire period of the 1920s can be categorized as an "uneasy" peace, the integration of Germany into the League and the Locarno agreements, as well as the recognition of the Soviet Union by the European powers presented some level of stability. See McKercher, 176.

24. For the text and analysis of the Draft Treaty, see Stone, "The Draft Treaty for the World Disarmament Conference."

25. On the work of the commissions, see Woodward, "Limitation of Land Armaments," and Woodward, "Limitation of Air Armaments."

26. Several scholars who have studies the history of the commissions and the development of the Geneva Conference have placed a considerable degree of the fault for failure of the Conference at the feet of the military-technical committees and the commissions. Though the records of these important dialogues were kept confidential and have not been made public, it seems clear from the results of the plenary sessions (which were public) that the military experts had an exceedingly difficult time in defining categories as aggressive or defensive and thus failed to reach consensus. More critical authors have charged that the military representatives effectively used the difficult nature of categorizing technologies and military applications to "bog the process down" in details, and thus avoid meaningful consensus. At any rate, it reflects the inherent uncertainty created by the dynamic nature of technology and the fears of those most sensitive to the potentially dramatic impacts of technology to place their own states' interests in possible jeopardy in a changed future environment. For discussion see Philip Noel-Baker, *The First World Disarmament Conference*, 3.

27. Unfortunately, this was not the case. In fact, one of the real problems of the conference was that too many issues were left to be settled without any preexisting concrete agreements developed during the preparatory phase. Thus, while the various representatives could agree to the principle of qualitative disarmament and its perceived benefits, the inevitable descent into details consistently frustrated efforts in Geneva. See Boggs, *Attempts to Define and Limit "Aggressive" Armament*, 39; and Wheeler-Bennett, *Pipe Dream of Peace*, 24–25.

28. See Marlies Ter Borg, "Reducing Offensive Capabilities: The Attempt of 1932," *Journal of Peace Research* 29, no. 2 (May 1992): 145–60; and Vladislav Kokoshin and Andrei Zubock, "Opportunities Missed in 1932?" *International Affairs* (February 1989): 112–21. On the Japanese intervention in Manchuria, see Sara R. Smith, *The Manchurian Crisis, A Tragedy in International Relations* (New York: Columbia University Press, 1948).

29. Disarmament received both broad and deep support from the citizens of most countries. Moreover, a burgeoning NGO community was increasingly active in pressing national government and League officials to address disarmament in comprehensive terms because of its perceived centrality to lasting peace. See, for example, Lorna Jaffe, "Abolishing War? Military Disarmament at the Paris Peace Conference," and Cecelia Lynch, "A Matter of Controversy: The Peace Movement and British Arms Policy in the Interwar Period," in B. J. C. McKercher, ed., *Arms Limitations and Disarmament, Restraints on War, 1899–1939*.

30. The depths of the Great Depression were being felt around the globe. See Charles Kindleberger, *The World in Depression, 1929–1939* (Berkeley: University of California Press, 1986); and P. M. H. Bell, *The Origins of the Second World War in Europe*, chaps. 9 and 10.

31. There is clear evidence that Bruning saw the Geneva Conference as, quite likely, his last chance to deliver a political victory to the German people and thus save his government from a likely reactionary backlash. Moreover, other representatives to the Congress were both aware and sensitive to this dynamic. See Mildred S. Wertheimer, "Hitler and the German Political Crisis, 1932–1933," *Foreign Policy Reports* 8, no. 26 (XXXX):307–18.

32. Soviet diplomacy continued to suffer from its perceived commitment to revolution, and Stalin's consolidation of power (particularly the marginalization and eventual liquidation of more revolutionary elements of the leadership) did little to change the situation. However, Maxim Litvinov's constructive participation at the conference continues to be considered central to its "near success." See Stone, "The World Disarmament Conference, First Stage, February 2–March 17, 1932."

33. See Williamson Murray, "May 1940: Contingency and Fragility of the German RMA," in *The Dynamics of Military Revolution 1300–2050*, eds. MacGregor Know and Williamson Murray (Cambridge: Cambridge University Press, 2001), 154–74; Williamson Murray and Alan R. Millett, eds., *Military Innovation in the Interwar Period* (Cambridge: Cambridge University Press, 1996); James S. Corum, *The Roots of Blitzkrieg: Hans von Seeket and German Military Reform* (Lawrence: University Press of Kansas, 1992).

34. Boggs, *Attempts t o Define and Limit "Aggressive " Armament*, 91.

35. On the development of the tank, see Patrick Wright, *Tank* (New York: Viking, 2000).

36. Wright, *Tank*, 17.

37. Kristafer Ailslieger, "The Battle of Cambrai," *Armor* (September–October 2000), Travers, "Could the Tanks of 1918 Have Been War-Winners for the British Expeditionary Force?"; Bailey, "The First World War and the Birth of Modern Warfare," 140.

38. Boggs, *Attempts to Define and Limit "Aggressive" Armament*, 85.

39. Roger Edwards, *Panzer: A Revolution in Warfare, 1939–1945* (London: Brockhampton Press, 1989).

40. Williamson Murray, "Armored Warfare: The British, French, and German experiences," in *Military Innovation in the Interwar Period*, eds., Murray and Millett, 6–49.

41. Trevor Wilson and Robin Prior, "Conflict, Technology and the Impact of Industrialization: The Great War 1914–19," *Journal of Strategic Studies* 24, no. 3 (September 2001):128–57.

42. Edward P. Warner, "Can Aircraft Be Limited?" *Foreign Affairs* 10, no. 3 (April 1932):431.

43. Boggs, *Attempts to Define and Limit "Aggressive" Armament* , 87.

44. Williamson Murray, "Strategic bombing: The British, American, and German experiences," in *Military Innovation in the Interwar Period*, eds., Murray and Millet, 96–143.

45. Boggs, *Attempts to Define and Limit "Aggressive" Armament* , 87.

46. Both men were prolific writers. For a discussion of their works, see, for example, Brian Holden Reid, *Studies in British Military Thought: Debates with Fuller and Liddell Hart* (Lincoln: University of Nebraska Press, 1998).

47. See Corum, *The Roots of Blitzkreig*, Heinz Guderian, *Achtung-Panzer! The Development of Tank Warfare* (London: Cassell, 2000); Heinz Guderian, *Panzer Lead*er (New York: De Capo Press, 2002).

48. Sally W. Stoecker, Forging Stalin's Army, *Marshal Tukhachevsky and the Politics of Military Innovation* (Boulder, CO: Westview, 1998); David R. Stone, *Hammer and Rifle: The Militarization of the Soviet Union, 1926–1933* (Lawrence: University Press of Kansas, 2000); David M. Glantz, *Stumbling Colossus : The Red Army on the Eve of World War* (Lawrence: University Press of Kansas, 1998).

49. Richard Simpkin, *Deep Battle: The Brainchild of Marshal Tukhachevskii* (London: Brassey's, 1987); Alexi Kokoshin, *Soviet Strategic Thought, 1917–1991* (Cambridge: MIT Press, 1998).

50. Alan Clark, *Barbarossa: The Russian-German Conflict, 1941–1945* (New York: William Morrow, 1965); John Erickson, *The Road to Stalingrad:: Stalin's War with Germany Vol. 1* (New York: Harper and Row, 1975); David M. Glantz, *Barbarossa: Hitler's Invasion of Russia, 1941* (Charleston: Tempus Publishing, 2001).

51. Azar Gat. "Ideology, National Policy, Technology and Strategic Doctrine between the World Wars," *Journal of Strategic Studies* 24, no. 3 (September 2001): 5. Gat goes on to note that "'Deep Battle' was the most advanced and comprehensive operational doctrine developed in the 1930s" (p. 6), making the point that while the Blitzkrieg was certainly effective in World War II, it was the combination on the battlefield of several well-developed tactical executions, it was not a military doctrine.

52. See, for example, David MacIsaac, Voices from the Central Blue: The Air Power Theorists, in Paret, ed., *Makers of Modern Strategy*, 624–47.

53. Boggs, *Attempts to Define and Limit "Aggressive" Armament*, 77.

54. Edward P. Warner, "Aerial Armament and Disarmament," *Foreign Affairs* 4, no. 4 (1926): 624–36.

55. Ibid., 627.

56. Edward P. Warner, "Can Aircraft Be Limited?" *Foreign Affairs* 10, no. 3 (1932): 431–43.

57. Warner, "Aerial Armament and Disarmament," 628.

58. Warner, "Can Aircraft Be Limited?" 436–39.

59. Warner, "Aerial Armament and Disarmament," 633.

60. See, for example, Vittorio Scialoja, "Obstacles to Disarmament," *Foreign Affairs* 10, no. 2 (1932): 212–19; Alan Dulles, "Progress toward Disarmament," *Foreign Affairs* 11, no. 1 (1932): 54–65.

61. For discussions of early League disarmament activities, see Wertheimer, "The League of Nations and the Prevention of War."

62. Disarmament received a prominent place within the League framework, initially laid out in Article 8. The presence of large standing armies and commitments to large military industries was viewed as an important impediment to lasting peace and stability and a contributing factor to World War I. See Howard, *The First World War*, 22–25.

63. Quoted in Wheeler-Bennet's *Disarmament and Security Since Locarno, 1925–1931*, 25. This language was codified in Article 7 of the League Charter. See Wertheimer, "The League of Nations and the Prevention of War."

64. For a discussion of the Versailles settlement and the obligations on Germany and its allies, see Wheeler-Bennett, *Disarmament and Security Since Locarno, 1925–1931.*

65. The "battle for equality" between the disarmed Germany and its neighbors, particularly France and the states of the "Little Entente" centering on the need for the latter states to live up to their own obligations to disarm would eventually dominate the Geneva Conference. For discussion of this, see Wheeler-Bennett, *The Pipe Dream of Peace*, chap. 5.

66. Attempts to "size" the appropriate levels of forces or "effectives," for given states was also a difficult exercise. Given the differing interests and challenges of states (from internal security needs to policing of colonial territories), quantitative approaches should not be considered any more successful than qualitative ones. For a discussion of the attempts to regulate force structures, see "The Disarmament Deadlock."

67. This is not to say that informed military and political leaders saw the balance as necessarily shifting dramatically to offense. Defensive countermeasures would likely follow from any advances in offensive technology. It is only to say that many leading thinkers of the day saw the increased potential for aggression, and consequently, some of them, such a Basil Liddell Hart, argued that the abolition or severe limitation of these technologies could sustain the defensive character of the postwar system. For a discussion of this, see Azar Gat, "Ideology, National Policy, Technology and Strategic Doctrine between the World Wars," *Journal of Strategic Studies* 24, no. 3 (September 2000): 1–18.

68. The increased pressures of an offense-dominant system, where potentially decisive first-strikes are possible, hinders diplomacy and undermines stability. Stephen Van Evera makes this argument in chapter 3 of his book, *Causes of War*.

69. See, for example, George Quester, *Offense and Defense in the International System* (New York: Wiley, 1977); and Jervis, "Cooperation under the Security Dilemma."

70. See Boggs, *Attempts to Define and Limit "Aggressive" Armament*.

71. Boggs, *Attempts to Define and Limit "Aggressive" Armament*, 14.

72. See "The Disarmament Deadlock," 383.

73. Ibid, 383.

74. German frustrations throughout the period increased after their entry into the League and the signing of the Locarno agreements. Despite their efforts to be constructive and live up to their obligations, German diplomatic initiatives were consistently rebuffed by the French. See, for example, Stone, "The World Disarmament Conference, First Stage, February 2–March 17,

1932," 63; and Stone, "The World Disarmament Conference, Second Stage, March 17, 1932–January 1933," 272–73.

75. See, for example, Stone, "The World Disarmament Conference, First Stage, February 2 –March 17, 1932," 63–64; and Stone, "The World Disarmament Conference, Second Stage, March 17, 1932–January 1933," 271.

76. For an excellent discussion, see Melvyn P. Leffler, *The Elusive Quest: America's Pursuit of European Stability and French Security, 1919–1933* (Chapel Hill: University of North Carolina Press, 1979).

77. On deterrence, see John J. Mearsheimer, *Conventional Deterrence* (Ithaca, NY: Cornell University Press, 1983.

78. We can also see in France's strategy the type of sensitivity to relative power shifts central to the theory of dynamic differentials as discussed by Dale Copeland in *Origins of Major Wars*, chapter 2.

79. Peter Jackson, "France and the Problems of Security and International Disarmament after the First World War," *Journal of Strategic Studies* 29, vol. 2 (2006): 254.

80. For discussion of deterrent versus spiral model interactions, see Robert Jervis, *Perception and Misperception in International Politics* (Princeton, NJ: Princeton University Press, 1978).

81. The equality debate is discussed in detail by John W. Wheeler-Bennett, *The Pipe Dream of Peace, The Story of the Collapse of Disarmament* (New York, William Morrow, 1935), chaps. 5, 9.

82. On signaling, see Jervis, "Cooperation under the Security Dilemma"; Glaser, "Realists as Optimists"; and Andrew Kydd, "Sheep in Sheep's Clothing: Why Security Seekers Do Not Fight Each Other," *Security Studies* 7, no. 1 (Autumn 1997): 114–55.

83. On the internal German political situation, see Wertheimer, "Hitler and the German Political Crisis, 1932–1933."

84. Both the United States and Great Britain desired to see a stable and peaceful Germany fully integrated into a stable European system. France, however, saw security against a renewed German threat as more important than stability. See Jaffe, "Abolishing War?" 45.

85. Boggs, *Attempts to Define and Limit "Aggressive" Armament*, 34.

86. See Henry R. Winkler, "Arthur Henderson" in *The Diplomats 1919–1939*, eds. Gordon A. Craig and Felix Gilbert (Princeton, NJ: Princeton University Press, 1953), 341.

87. Dick Richardon, "The Geneva Disarmament Conference, 1932–1934," in *Decisions and Diplomacy: Essays in Twentieth Century International History*, eds. Dick Richardson and Glyn Stone (New York: Routledge, 1995), 68.

88. For a discussion, see Carolyn J. Kitching, *Britain a nd t he Geneva Disarmament Conference* (New York: Palgrave Macmillan, 2003).

89. Boggs, *Attempts to Define and Limit "Aggressive" Armament*, 30.

90. The speech was extremely well received by the publics of various states, but far less enthusiastically by leaders and diplomats in attendance.

91. See Henry L. Roberts, "Maxim Litvinov," in Gordon A. Craig and Felix Gilbert, *The Diplomats 1919–1939* (Princeton, NJ: Princeton University Press, 1953), 351. Litvinov, however, still challenged the possibility of improving international relations except by the path of disarmament. On February 11, 1932, at the Disarmament Conference he declared: "Security against war must be created. This security can never be achieved by roundabout ways, but only by the direct way of total general disarmament. This is no communist slogan. The Soviet delegation knows that the triumph socialistic principles, removing the causes giving rise to armed conflicts, is the only absolute guarantee of peace. So long, however, as these principles prevail only in one-sixth of the world, there is only one means of organizing security against war, and that is total and general disarmament." A year later, on February 6, 1933—that is, after Hitler had come to power—Litvinov took a different line. He expressed, somewhat reluctantly, his willingness to consider the French security proposal, and in conjunction with a French plan submitted a draft definition of aggression. He was now attempting to supply criteria which in 1928 he had said were lacking.

92. See H. Stuart Hughes, "The Early Diplomacy of Italian Fascism, 1922–1932," in *The Diplomats 1919–1939*, Gordon A. Craig and Felix Gilbert (Princeton, NJ: Princeton University

Press, 1953), 230. "And so for a certain time in Italian foreign relations the traditions of the old diplomacy and the impatience of the new came to a kind of happy compromise. But this was a special and peculiar case. It was because Italy was a lesser power and most of the international issues in which she was involved were relatively unimportant. During the 1920s, after all, it was France and Britain that were really making the great decisions. Italy remained on the fringes of major diplomacy. So long as she stayed there, the divergences of view between Mussolini and his professional advisers could be compromised or glossed over. But once there should come a radical change in European alignments and Italy could succeed in forcing her way into the main stream of international negotiation, the fragile compromise between Fascism and tradition could scarcely fail to collapse.

93. B. J. C. McKercher, "Of Horns and Teeth: The Preparatory Commission and the World Disarmament Conference, 1926–1934," in *Arms Limitation and Disarmament, Restraints on War, 1899–1939*, ed. McKercher (Westport, CT: Praeger, 1992), 183–84.

94. For additional discussion and analysis of the French plans and their focus on internationalization, see Boggs, *Attempts to Define and Limit "Aggressive" Armament*, 25–30.

95. Boggs, *Attempts to Define and Limit "Aggressive" Armament*, 27–28.

96. Boggs, *Attempts to Define and Limit "Aggressive" Armament*, 28. Also quoted in Borg, "Reducing Offensive Capabilities," 156.

97. Philip Noel-Baker, *The First World Disarmament Conference 1932–33 a nd Why I t Failed.*

98. Boggs, *Attempts to Define and Limit "Aggressive" Armament*, 82–83.

99. Quoted in Boggs, *Attempts to Define and Limit "Aggressive" Armament*, 81.

100. Marlies Ter Borg, "Reducing Offensive Capabilities," 149.

101. As Zubok and Kokoshin point out: "technical subcommittees were set up, with professional military experts making up two-thirds of them. They met behind closed doors so as not to attract public attentions, nor did they bring independent government experts and scientists into discussing the problems that cropped up. This enabled military officials of Britain, France, the United States, Japan, and other capitalist countries to drown the substance of the matter in numerous reservations and technical details. Their findings implied that virtually all weapons in which the great powers were superior were purely defensive and presented no great menace to other countries" (p. 119).

102. Zubok and Kokoshin, p. 114.

103. McKercher, *Arms Limitation and Disarmament*, xiii.

104. Quoted in Phillip Noel Baker, *The First World Disarmament Conference 1932–33 a nd Why I t Failed*, 78.

105. Quoted in Wheeler-Bennett, *The Pipe Dream of Peace*, 25, fn. 1.

106. Ibid., 16–20.

107. Ibid., 28.

108. Ibid.

109. Ibid., 29.

110. Ibid., 29–30.

111. Ibid., 31–34.

112. Ibid., 32.

113. Ibid., 34.

114. For details of the Hoover plan, see Stone, *World Disarmament Conference, Second Stage*, 277–78.

115. McKercher, *Arms Limitation and Disarmament*, 186.

116. In discussions of naval armaments, the U.S. defined all ships except for submarines as essentially defensive. This included aircraft carriers. Since the 1920s, the United States had experimented with carrier aviation, based on the perceived need of projecting force into the western Pacific in the event of a future war with Japan, or at the very least to safeguard its perceived interest in an "open door" to China. Given the objective nature of the missions of the aircraft carrier, premised on projecting power over long ranges (certainly longer than any battleship could project), it is difficult to make the argument that they were defensive in nature. However, after the 1920 Washington Conference, which set ratios for the U.S., Japanese, and British navies and also limited both U.S. and British island fortifications to allow the Japanese

predominance in the western Pacific, the perceived need to project power over long-ranges of ocean became central to U.S. planning. For the Japanese, who clearly wanted to avoid U.S. penetration into its perceived sphere of influence, the aircraft carrier could only be seen as offensive, and thus something that would be better banned. It is ironic, of course, that the Japanese used their aircraft carrier fleet to launch a devastating first strike on the U.S. fleet at Pearl Harbor.

117. McKercher, *Arms Limitation and Disarmament*, 182.

118. McKercher, *Arms Limitation and Disarmament*, 185.

119. Boggs, *Attempts to Define and Limit "Aggressive" Armament*, 33.

120. Stone, *The World Disarmament Conference, Second Stage*, 274–76.

121. See Boggs, *Attempts to Define and Limit "Aggressive" Armament*, 32–33.

122. Kitching, *Britain and the Geneva Disarmament Conference*, 98–102.

123. Ibid., 130–35.

124. Ibid., 154.

125. Kitching, *Britain and the Geneva Disarmament Conference*, 174. Also see Wheeler-Bennett, *Pipe Dream of Peace*, 180–81.

126. McKercher, *Arms Limitation and Disarmament*, 193.

127. Craig, *The Diplomats 1919–1939*, 233. "Grandi was becoming a feeble support to lean on. He also was beginning to feel the ground slipping from beneath his feet. Back in March 1930, at the time of the London Naval Conference, he had confessed to Prime Minister MacDonald that he detected in a recent declaration by the Grand Council of Fascism 'a veiled admonition against himself for having gone as far as he had' toward a compromise with the French. The following June he had complained to the French Ambassador of the activities of the 'strong party . . . which wished to throw Italy into the arms of Germany' and had characterized his own position as 'one of extreme difficulty.' Part of the lament, of course, had been for effect—to induce the French to come closer to the Italian viewpoint. But the distress had been real. . . . On June 16, 1932—just as the Conference on Disarmament at Geneva was beginning to adjourn its inconclusive labors—Mussolini gave an interview to the *Daily Express*, in which he voiced his pessimism about the League of Nations and about the Conference itself. To the initiated it was a clear indication that Grandi's time had come" (p.).

128. The fact that so much of the technical and military "details" had been left to be determined at the conference did not help. However, the "battle for equality" repeatedly undermined the type of consensus needed to truly grapple with the harder issues of qualitative disarmament.

129. Jervis, "Cooperation under the Security Dilemma."

130. E. H. Carr, *The Twenty Years Crisis, 1919–1939: An Introduction to the Study of International Relations* (New York: Harper and Row, 1964), 74.

131. For this insight, I appreciate the suggestion of Dr. Len Schopa, University of Virginia.

132. Thomas C. Schelling, *Arms and Influence* (New Haven, CT: Yale University Press, 1966), 249.

133. Schelling, makes this point, see page 248.

134. See, for example, *Raising Churchill's Army: The British Army and the War against Germany 1919–1945* (Oxford: Oxford University Press, 2000).

Chapter 4

The Baruch Plan and the Atomic Bomb

There can be little doubt that the atomic bomb was viewed as a decisive military application by those responsible for its eventual use.[1] While the initially small number of usable weapons limited the potential utility of the bomb in U.S. strategic planning in the period of time leading up to the Japanese surrender and the immediate aftermath, the larger ramifications of the bomb penetrated all aspects of U.S. political and military leadership. Soon the perceived benefits of a tough, hardline "atomic diplomacy" initially advocated by Secretary James Byrnes in various postwar negotiations quickly evaporated.[2] Among U.S. leaders, the decisive nature of the atomic bomb created a stark new reality. This led to two seemingly paradoxical lines of thought. First, the atomic bomb represented a truly revolutionary weapon capable of destroying mankind, but so long as it was only possessed by one state, others would almost certainly strive to obtain it. Second, because the bomb was so dangerous, particularly to an advanced society like the United States, with large urban population centers, its possession by any other power would be an unacceptable threat to the survival of the United States. Despite its clear lead in atomic weaponry, the threat associated with proliferation of atomic weapons would undermine any temporary advantage held by the United States. These two implications drove U.S. planners to a somewhat unique conclusion: effective international control of atomic energy—both its military and potential vast civilian applications—seemed to provide the only potential solution to the fundamental security challenge created by the atomic bomb. If the victorious powers could agree to the development of an international organization that could effectively control atomic energy (and thus atomic weapons), then this horrific outcome could be avoided.

However, precisely because of the decisiveness of the bomb, the United States had to maintain its "monopoly" so long as any question remained about the effectiveness of international control. The absolute nature of the weapon created a perceived need for absolute security for those who created it. Thus, while the United States' diplomatic position on atomic energy was that international control was a singular priority, central to peace and security in the postwar world, the United States also refused to accept anything less than a "fool-proof" system that could guarantee its security. Without such a system it was unwilling and unable to countenance limitations on its nuclear stockpile. While the U.S.-Soviet relationship deteriorated from a war time alliance over the course of discussions to Cold War rivalry over a host of important geopolitical issues, the control of atomic energy was seen as fundamentally important and in some ways distinct from other disputes. While other theoretical approaches may explain the ultimate failure of international control, MET explains both the motivations behind the U.S. offer as well as the Soviet rejection. Moscow could never agree to forego such a decisive weapon precisely because of its perceived revolutionary nature and capabilities. Only its possession could begin to approach a fulfillment of those needs. Unsurprisingly, not even a revolutionary, highly ambitious and technically comprehensive system of control could effectively address the deep insecurity instilled in both states by this weapon.

This chapter proceeds as follows. The first section will consider the postwar security environment. Despite the eventual onset of the Cold War rivalry, the political-military environment during the period from 1945–1947 remained uncertain. The second section will examine the views of civilian and military leaders concerning the impact of atomic weapons in the United States, while the third section will consider the views of Stalin and his advisors. The fourth section will consider the so-called Baruch Plan and its predecessor the Acheson-Lilienthal plan. The efforts to develop a workable, functioning, and credible system to effectively monitor the international control of atomic energy were impressive. Whether with the benefit of hindsight they now also look naïve or extremely optimistic should not diminish the gravity with which the task was undertaken. The fifth section will revisit the history of the negotiations of the United Nations Atomic Energy Commission (UNAEC) and its failure to adopt the Baruch Plan or a possible Soviet counterproposal. The conclusion will consider this case and the respective explanations of MET as well as other alternative hypotheses.

THE EMERGING SUPERPOWER SECURITY ENVIRONMENT

While U.S. and Soviet leaders faced difficult challenges in constructing a postwar international order, they were engaged in cooperative negotiations on issues. The potential control of atomic energy was an area that seemed viable for international cooperation under the auspices of the United National

Security Council. In preparing to approach the Soviets, the first task was to foster agreement between the three nations that had contributed to building the bomb, the United States, Great Britain, and Canada.[3] The proposal laid out the primary objectives of the commission including the exchange of basic scientific information for peaceful exploitation of atomic energy, the control of atomic energy to ensure its use for peaceful purposes, the elimination of atomic weapons from national arsenals, and effective safeguards against violations and evasions. The United States, Great Britain, and the Soviet Union, with France, China, and Canada, would cosponsor a resolution to establish a United Nations Atomic Energy Commission (UNAEC) at the inaugural session of the UN General Assembly planned for January 1946.[4]

Secretary of State Byrnes's trip to Moscow was perceived to be a test for how receptive the Soviets would be to the issue of international control.[5] The Soviets had remained relatively quiet on the issue of atomic energy until early November when Foreign Minister Molotov stated before a Kremlin audience that "it is not possible at the present time for a technical secret of any great size to remain the exclusive possession of one country," and more ominously that "we shall make up properly for all lost time. . . . We shall have atomic energy and many other things too."[6] Reports from U.S. and British diplomats in the Soviet Union evaluating the impact of the atomic bomb on Soviet policy had concluded that Soviet cooperation might be difficult to obtain precisely because the Soviets had been shocked by the weapons and would only be satisfied with its acquisition.[7]

Aside from an initial argument by Molotov to address the issue of atomic energy last, rather than first, as Byrnes had suggested, the Soviets accepted the proposal with little argument on December 26.[8] The only major Soviet change to the agreement was an explicit recognition of the priority of the UN Security Council over the planned commission, thus protecting the Soviet veto power. Byrnes and British Foreign Minister Bevin agreed to this adjustment. Prior to leaving Moscow, Byrnes announced the success of the meetings without first communicating the results to Washington. Without details on the agreement, skeptics were infuriated that Byrnes might have indeed crafted a bargain with the Soviets that would lock the United States into sharing vital information without safeguards or compromised on other issues. The agreement paved the way for work to begin on the Commission and seemed to (at least temporarily) create a pathway for cooperation on the critical issue while dampening perceptions of Soviet intransigence. Unfortunately, this would indeed prove a temporary lull in the diplomatic battle over international control and soon President Truman's heated lecture to Byrnes that he was "tired of babying the Soviets" would prove well-founded.[9]

The relatively positive feelings created in Moscow quickly evaporated at the outset of the opening session of the General Assembly in London in January 1946. Various procedural issues became bogged down in acrimony.

However, even against this rather contentious backdrop, the General Assembly unanimously passed as its first resolution the creation of a UNAEC, based on the verbatim text of the Moscow Declaration, on January 24.[10] Throughout this process, and despite the concerns about Byrnes's handling of the matter with the Soviets, U.S. policy remained clear and unchanged: the U.S. atomic bomb program would remain in operation until such time as an international control agency, imbued with the necessary authority to institute rigorous safeguards against any possible transgressions by other states, was fully operational. The fear of an atomic arms race and a possible atomic war drove the search for international control, but fear of anything other than an airtight guarantee against atomic weapons in the possession of other states (most importantly the Soviet Union) drove the approach to achieving that control.

During this period, bilateral relations between the United States and the Soviet Union deteriorated further. In mid-July 1946 the United States carried out a series of atomic tests on the Bikini Atoll in the South Pacific.[11] Predictably, this test, undertaken less than a month after the initial meeting of the UNAEC, and during the period of seemingly in-depth negotiations created an uproar. It provided a propaganda opportunity for the Soviets and also undermined the credibility of the United States in terms of its stated commitment to international control. Important evaluations of Soviet behavior and intentions such as George Kennan's famous "Long Telegram" and the Clifford-Elsey Report had been disseminated throughout the Truman administration and the view of the key policy makers within the administration had shifted from suspicion to hostility. Still, this did not mean that international control of atomic energy was impossible; instead the administration was resigned to a longer process, and was, to a significant degree, willing to delink the process of the UNAEC from other bilateral relations.[12]

THE U.S. VIEWS ON THE ATOMIC BOMB

Even before the atomic bombs were dropped on Hiroshima and Nagasaki, the magnitude of its impact on both military affairs and international relations was being discussed and debated by those who created it. Niels Bohr made a personal appeal to President Roosevelt on August 6, 1944, and presented a memorandum that outlined his fears about the prospects of an atomic arms race and its frightening implications.[13] Similarly, Leo Slizard attempted to meet with Roosevelt before he died, and ultimately presented a petition signed by 59 of his colleagues to President Truman in July 1945.[14] Slizard was also a member of the committee that drafted the Franck Report. Presented in June 1945 by the leading scientists of the Manhattan Project's Metallurgical Laboratory at the University of Chicago, headed by Dr. James Franck, the report articulated a primary concern about the nature of the technology and materiel needed to successfully develop an atomic bomb program.[15] They argued that neither the know-how, nor the raw materials (uranium and thorium) could be successfully monopolized. The effective control over these two essential components was necessary for any "atomic monopoly" to be sustained for any meaningful period of time. "Thus we cannot hope to avoid a nuclear armament race either by keeping secret from the competing nations the basic scientific facts of nuclear power or by cornering the raw materials required for such a race."[16]

Ultimately these recommendations failed to restrain the United States from utilizing the atomic bomb to end the Pacific War, but the broad characteristics of the solution developed by the members of the Franck Panel would remain the central concepts of the U.S. strategy to control atomic energy in the postwar period.[17] The harsh new reality was that there was little that any state, including a great power like the United States, could do to defend itself against the devastation of an atomic attack. Unilateral means of protection were seemingly nonexistent. Thus, to effectively address the challenge of the bomb and the larger issue of atomic energy, states would have to work together and cooperate to create a system capable of maintaining international security and removing the fears and temptations of atomic weapons.[18]

Yet as much as the remedy—international control—seemed to be the only real strategy, the decisive nature of these weapons also shaped exactly how far the U.S. government could actually go in its pursuit of developing and creating a cooperative international solution to this fundamental security challenge. The only solution was complete and effective international control. This would require the development of an international organization that had broad and intrusive powers which would likely infringe upon, if not override, traditional notions of state sovereignty. "One thing is clear: any international agreement on prevention of nuclear armaments must be backed by actual and efficient controls. No paper agreement can be sufficient, since neither this or any other nation can stake its whole existence on trust in other nation's signatures."[19]

International control may have been the preferred answer, but the nature of the weapon demanded that the control be complete and unquestionable. This would be a standard that would repeatedly shape and hinder the political and diplomatic discussions on the subject. Only with such a structure could the absolute security from the absolute weapon be guaranteed. While the argument had failed to restrain the United States from utilizing the atomic bomb against Japan, the postwar environment seemed to present a new opportunity for international control and the development of robust means to maintain and enforce that control. While relations with Soviet Russia were perceived as highly uncertain, many in the Truman administration believed that a real opportunity to develop this type of international control existed, particularly if the United States was able to approach the Soviets in a way that fostered cooperation rather than coercion.

The Views of Truman's Civilian Advisors

It was with this and other similar recommendations in mind that one of the key political figures in the development of the atomic bomb attempted to advocate for the views of the scientists to other key members of the administration.[20] Secretary of Defense Henry Stimson advised President Truman in September 1945 about the importance of approaching the Soviets directly about developing a postwar regime to control atomic energy. Stimson felt strongly that the atomic bomb was indeed a revolutionary weapon and that the Soviets would inevitably seek to obtain it as soon as possible. "I think the bomb instead constitutes merely a first step in a new control by man over the forces of nature too revolutionary and dangerous to fit into the old concepts."[21] Moreover, he believed that the means of addressing the issue was to deal with our ally directly (in consultation with the British), reflecting the seriousness of the U.S. commitment to a cooperative solution and avoiding the potential confusion and muddling of the issue in an open forum like the proposed United Nations.[22] "It is my judgment that the Soviet would be more apt to respond sincerely to a direct and forthright approach made by the United States on this subject than would be the case if the approach were made as a part of a general international scheme, or if the approach were made after a succession of express or implied threats or near threats in our peace negotiations."[23] It was Stimson's fear that this was precisely the approach that Secretary of State Byrnes and President Truman had envisioned in their negotiations with Stalin.[24]

So far, the Soviets had exhibited little response, positive or negative, to the existence of the U.S. atomic weapon. Truman's offhanded announcement of the bomb to Stalin at the Potsdam conference had elicited little response. As Truman later wrote, "On July 24 I casually mentioned to Stalin that we had a new weapon of unusual destructive force. The Russian Premier showed no special interest. All he said was he was glad to hear it and hoped we would make 'good use of it against the Japanese.'"[25] Others who were privy to the conversation, including Secretary of State James Byrnes seemed to agree that Stalin didn't understand the gravity of the announcement. Byrnes felt that "he had not grasped the importance of the discovery."[26] As will be discussed below, Stalin's feigned lack of interest was merely a tactic to deliberately downplay the importance of the bomb. Understanding well the nature of the atomic weapon, he feared that recognition of the new U.S. advantage would only contribute to the utilization of the weapon as diplomatic leverage by the Americans in ongoing negotiations. It seems that he was not only well aware of the U.S. progress on the atomic bomb project, but was also deeply concerned about its implications for the burgeoning U.S.-Soviet relationship. His response would be to channel all available resources into a Soviet atomic program that would rectify this dangerous imbalance.[27]

Stimson saw a window of opportunity to draw the Soviets into an agreement on atomic energy, based on their interests and concerns. Moreover, he saw that it was critical to focus on the issue of atomic energy and avoid attempts to use the bomb as leverage against the Soviets relative to their behavior in Europe or in any attempt to "force open" Soviet society as some in the administration were already arguing.[28] It was imperative, in Stimson's view, that the Soviets be brought into an arrangement for the control of atomic energy as willing partners without being coerced by their former allies. While the discovery of the bomb provided America with a real advantage, the twin acknowledgments that this advantage would not last and that an eventual Soviet atomic bomb would likely mean the creation of an inherently dangerous and unstable world system demanded U.S. leadership. However, this leadership had to be premised on a realistic basis that acknowledged Soviet interests, including their commitment to state control within the Soviet Union itself.[29]

Stimson's role in the development of the atomic bomb coupled with his respected position within the Truman administration granted him the influence to initiate a major assessment of official U.S. policies with regard to atomic weapons as well as the larger question of atomic energy. In an important September 25 meeting, Truman assembled a number of his top advisors to attempt to generate a consensus on how the government should address the issue and its international and domestic implications. Two of Stimson's most vehement supporters were Undersecretary of State Dean Acheson, and Stimson's undersecretary and successor in the War Department, Robert Patterson.

Acheson accepted Stimson's argument as the most logical and effective premise for U.S. policy.[30] Most importantly, he argued "a policy of secrecy is both futile and dangerous and that the real issues involved the methods and conditions which should govern interchange of scientific knowledge and the international controls which should be sought to prevent a race toward mutual destruction."[31] In setting out the details of his own position, Acheson argued that several important facts should be taken as the assumptions of how the United States should approach the question of international control. The key points were that the atomic "secret" was not in fact a secret and that the administration should be fully aware that scientific knowledge necessary to develop atomic weapons was already "widely known." Thus, there was little or no reason to believe that other states would not be able to catch up to the United States in fairly short order. Instead, the United States should take as granted "that foreign research can come abreast of our present knowledge in a comparatively short time," and therefore, "that foreign industrial engineering and development in, for instance, the Soviet Union can equal our present development in about five years."[32] This position was only reinforced by a more general and far-reaching understanding of what the discovery of atomic energy meant in larger scientific and historical terms. "The scientific knowledge does not relate merely to another and more powerful weapon. It relates to a discovery more revolutionary in human society than the invention of the wheel, the use of metals, or the steam or internal combustion engine. Its development cannot be suppressed or confined to one country."[33]

Thus the combination of the inevitability of technological advance in the field of atomic energy, the perceived weakness of the United States in the face of a possible nuclear adversary, and the certainty of Soviet acquisition of atomic weapons left international control as the only real option to be pursued. It was already possible to see what would happen in the event that no international control could be developed. A "head-start" meant little. "In other words, if the invention is developed and used destructively there will be no victor and there may be no civilization remaining. The advantage of being ahead in such a race is nothing compared with not having the race."[34]

Acheson argued that the only strategy that held any prospect of maintaining U.S. security was to reach out to the Soviet Union and develop a cooperative approach to dealing with the serious challenges that atomic weapons posed to both states and to international security more broadly. He did not have any illusions that the task would be easy or that negotiating with the Soviets in the context of the larger diplomatic environment would immediately yield results, but it was the only prudent course.[35] Secretary Forrestal would consistently articulate a more hawkish position on the issue of international control and of cooperation with the Soviet Union more generally. As Secretary of the Navy and as the eventual first Secretary of Defense, he would often take a harder line than even his military subordinates. Like his colleagues, he clearly saw the bomb as a revolutionary weapon, but he refused to believe that international cooperation could ever succeed to such an extent that it would actually safeguard U.S. security interests. Only through the continuing development and deployment of nuclear weapons in the larger context of a strong U.S. military could America achieve any real security against an inevitable Soviet threat. Maintaining the current advantage and keeping the edge was the preferred policy even if the danger of atomic war increased. As the October 10 meeting minutes indicate, Forrestal was against turning the bomb over to "a piece of paper."[36] This concern was not ignored. In fact, the views of Forrestal and the military would continue to play a role in shaping exactly what the eventual American proposal for international control would look like, and the levels of safeguards necessary that would be required for U.S. cooperation. It was with these realistic concerns, rather than any naïve desires for multilateral cooperative solutions, that international control of atomic energy was made a priority for U.S. security and a central issue of U.S. diplomacy.

Truman was convinced by the arguments of Stimson and others, though it had taken some time. While his initial views on the U.S. atomic monopoly had instilled a sense of confidence in dealing with the Soviet from a position of advantage or strength, the advocacy of Stimson and his supporters had tempered the president's enthusiasm.[37] However, the very magnitude of the discovery of atomic nature seemed to demand new thinking and new structures of control and governance to safely address the challenges it has created.[38] If the United States and other nations were unable to develop an effective system of international control, the failure would constitute not only a threat to American security, but to the world.[39]

The U.S. Military View of Atomic Weapons

Given his central role in the development of America's atomic weapon, it is unsurprising that General Lesley Groves would play a major role in thinking through the military implications of the new weapon for U.S. security. In his memorandum, "Our Army of the Future—As Influenced by Nuclear Weapons," Groves developed the central concepts that would shape the discussion of atomic weapons.[40] Given the nature of the atomic bomb, it was essential that the United States capitalize on its early advantage and maintain the edge it had developed. At the same time, he recognized the logical implications that followed from the likely diffusion of nuclear technology to other states. "Should there be an armament race in atomic weapons,—and that the world could not long survive such a race—then the United States must for all time maintain absolute supremacy in atomic weapons, including number, size and power, efficiency, means for immediate offensive use and defense against atomic attack."[41] In an atomic world, it was clear that there was little real (or perceived) defense against atomic weapons. States could dramatically "hurt" other states with this weapon, but could do little to prevent that harm in a world where other states possessed atomic arsenals.[42]

What was equally troubling was that while it was likely that the explosive power of atomic weapons could be increased in magnitude and scale, there was little evidence that any realistically effective countermeasures were in the offing over the foreseeable future. Improvements would only exacerbate this problem. The atomic bomb "can be simplified and improved from its early models and if we are to judge from past developments of other weapons future bombs should be more powerful than the bombs used at Hiroshima and Nagasaki."[43] Thus, military forces equipped with atomic weapons presented the potential for highly decisive offensive operations against adversaries, but defense against such attacks was essentially nonexistent.

The centrality of atomic weapons in this new era of political-military relations carried with it a new burden, that of developing and understanding the activities of other states with regard to atomic technology.[44] Only with a solid understanding of the capabilities and intentions of other states could the U.S. military and political leadership develop effective plans and actions to safeguard U.S. security. The United States would have to remain on high alert against potential attack from other atomic powers, in order to seize the initiative and, if necessary, preemptively strike its enemies. Specifically, Groves articulated the need for overseas bases for U.S. assets to respond quickly to potential aggression. The nature of atomic weapons would preclude the ability of time-consuming mobilizations. Envisioning the type of roles and missions that would soon be undertaken by the Strategic Air Command (SAC), he concluded that the U.S. military would have to maintain a high state of readiness and certain assets would likely be on a near constant state of high alert. Seizing the initiative and striking the first blow would be decisive in a future atomic war.[45] Groves also seemed well-aware of the inherent difficulties that American leaders would face in articulating, never mind carrying out, such a preventive strategy.

> If we were truly realistic instead of idealistic, as we appear to be, we would not permit any foreign power with which we are not firmly allied, and which we do not have absolute confidence to make or possess atomic weapons. If such a country started to make atomic weapons we would destroy its capacity to make them before it had progressed far enough to threaten us.[46]

This passage has been highlighted for laying the ground for eventual U.S. war plans focused on preventive attacks against the Soviet Union and criticized for its belligerence.[47] However, Groves was grappling with the same fundamental problem faced by the civilian decision makers who struggled with the issue of control of atomic energy during this period. Even with his more optimistic view on the perpetuation of the U.S. atomic monopoly, Groves understood that it was inevitable, at some point, that other states would reach an atomic breakthrough. As a military advisor, he was confronted with a grave threat that he deemed unlikely to be solved by international cooperation. Building upon a premise that cooperative initiatives to safely address the threat of atomic weapons and their proliferation were improbable, if not impossible, the gravity of the problem nonetheless demanded a solution. However unpleasant the prospect of preventive nuclear war to avert the development of nuclear weapons by a potential adversary, and understanding the potential destruction that could be visited upon the United States in the event of an atomic conflict, Groves articulated an approach that seemed to be the best of what were all bad options. Ultimately, this led Groves to a similar point that had been expressed by the scientists and civilian advisers, but with a greater emphasis on the necessary policies should cooperation fail.[48]

While he may have been skeptical of the prospects for international control, Groves concluded that the only feasible approach available was to support the diplomatic efforts to achieve it. However, while he contributed to the development of the U.S. plans to create an international body that could effectively manage this singularly important problem of American security, he also pushed for the development of military capabilities to provide the best chance to defend the security of the United States should international cooperation fail. This was not a contradictory position, and reflected the underlying tension that would define the American approach to control of atomic energy.

> If there were only some way to make America sense now its true peril some 15 to 20 years hence in a world of unrestricted atom bombs, the nation would rise up a demand one of two alternatives essential to its very existence. Either we must have a hard-boiled, realistic, enforceable, world-agreement ensuring the outlawing of atomic weapons or we and our dependable allies must have exclusive supremacy in the field, which means that no other nation can be permitted to have atomic weapons. The United States is n a position now to get and enforce one of those agreements.[49]

Like Truman's civilian advisors, Groves set a very high bar for the level of cooperation that would be necessary for the United States to rationally give up its arsenal. There could be absolutely no doubt about the intentions and capabilities of other states. Complete transparency at all times and everywhere would be the only foundation upon which a system of international control could be built. "This means the abandonment of all rights of privacy—that of the home, the laboratory and the industrial plant throughout the world."[50] While Groves and other military leaders would support the attempt to develop rigorous international controls over atomic energy, they would also be quick to remind the civilian leaders of the severe costs that would face the United States if a proposed system was anything less than foolproof. However, because of the consensus that developed even among the most ardent supporters of control, this argument typically was taken for granted. Given his central role in the development of atomic weapons, it is not surprising that Groves's views would be generally shared within the larger military establishment.[51]

This discussion underscores what would become the dual premises of U.S. policy in the construction of international control. The United States, as the first nation to successfully achieve nuclear fission, and seeking to avoid the dangers of an arms race would support the development of an international body that could aid the states of the world to harness the beneficial aspects of atomic energy, while effectively removing the dangers and insecurity created by atomic weapons. President Truman recognized the problems created by an international arms race, but also saw the promise of nuclear energy. Moreover, he saw the opportunity for U.S. leadership in the context of a United Nations to create a workable system that could address U.S. national interests while supporting international cooperation. While there were no guarantees that the Soviets would comply or support his initiative, it seems clear that President Truman sincerely believed that it was worth it to both the United States and the international community to make such an offer.[52]

THE SOVIET VIEW OF ATOMIC WEAPONS

It is clear that Soviet leader Josef Stalin understood the importance of the atomic bomb even before the Hiroshima bombing. President Truman interpreted Stalin's rather unassuming response to the revelation of a powerful new weapon in the U.S. arsenal as a lack of understanding of the significance of the weapon.[53] Secretary Byrnes agreed, particularly after Molotov showed little interest in the program in subsequent meetings. Nothing could have been further from the truth. Stalin understood exactly what this casual announcement by Truman meant for the Soviet Union and the postwar world.[54] The United States had obtained a decisive new capability, and while its existing stocks of atomic weapons might not be adequate to truly threaten the Soviet Union in the short term, the diplomatic objectives that Stalin sought in a postwar order were now in jeopardy. His reaction was swift and left little doubt about whether he considered atomic weapons important. Marshal Zhukov, who was in the room when Truman broke the news to Stalin, confirmed that his seeming lack of interest was a ruse. "In actual fact, on returning to his quarters after this meeting, Stalin, in my presence, told Molotov about his conversation with Truman. The latter reacted almost immediately. 'Let them. We'll have to talk it over with Kurchatov and get him to speed things up.'"[55]

Similarly, Gromyko, who was summoned to Stalin's quarters soon after the announcement also depicts a very different reaction from the one viewed the U.S. delegation.[56]

What this passage makes clear is that Stalin was deeply shaken, and disappointed with this dramatic turn of events. He seems to have been personally affected by the disclosure, and his remark about Roosevelt indicates a real sense of betrayal. Atomic energy was revolutionary and the development of an atomic weapon was a decisive achievement with enormous political and security implications. We can safely conclude that Stalin viewed the atomic bomb as a decisive weapon. It also seems clear that Stalin was unwilling to countenance any plan to forego the development of a Soviet bomb. If anything, the possibility of international control could be broached after the

Soviets had developed their own weapon, but the singular focus of the Soviet system from this point on would be to attain the atomic bomb.[57]

Thus Stalin's strategy for dealing with this troubling new situation contained two important components: a commitment to develop a Soviet bomb as soon as possible, and diplomatic strategy for dealing with the United States to counter their "atomic diplomacy." First, Stalin ordered the initiation of a crash program that would have all the resources deemed necessary to accomplish the goal of developing a Soviet atomic bomb. The existing bare-bones program under Kurchatov's direction would be transformed into a large-scale operation. The priority Stalin gave to developing the bomb is also reflected in the actions he took within the Soviet bureaucratic structure to achieve this objective.[58] Specifically, Stalin's appointment of Laventii Beria with the task of operating the atomic program reflected its priority.[59] Beria, arguably Stalin's closest confidant, had achieved his position and status within the leadership through ruthlessness and effectiveness. As the head of the NKVD and the GRU, Beria controlled one of the essential organs of the Soviet police state, including intelligence and security. His position would allow him to direct the necessary resources toward the program, working through (or, if necessary, around) the bureaucracy and using all tools, including coercion, to achieve the goal of a Soviet bomb.[60] As Kurchatov's notes from a meeting with Stalin, Beria and Molotov indicate, the bomb project was viewed as crucial to Soviet interests.[61]

In addition, the acquisition of resources from the occupied territories would fall under his direction, allowing the Soviet atomic program to benefit greatly from an infusion of German technology, as well as to gain intellectual and industrial capital. Stalin made sure to underscore this point. "It is also necessary to use Germany to the utmost; there, there are people, and equipment, and experience, and factories."[62] Lastly, Beria's access to information acquired through espionage would also contribute to the development program in the most efficient and effective ways given the constraints of Soviet bureaucracy. The Soviet program would prove remarkably successful, conducting an atomic test within five years in 1949.

THE ORIGINS OF THE BARUCH PLAN AND THE ATOMIC ENERGY COMMISSION

The Acheson-Lilienthal Report

Before heading to London for the opening session of the United Nations, Secretary Byrnes appointed Undersecretary Acheson to head a committee tasked with the development of a coherent U.S. policy on international control of weapons.[63] Their mission was to study "the controls and safeguards necessary" to protect the United States under an international system. Specifically, the committee would have to address the process of "stages" that would be required to achieve international control as put forth in the Moscow agreement. They would also examine the nature of the controls and inspection regime entrusted with the maintenance of a system. More broadly, the committee would assess the position of the United States with regard to atomic energy in the context of global technological development.

Early on in this process the views of the members coalesced around two important assumptions that would guide their collective thinking about the problem of control. The first was that atomic weapons were uniquely dangerous and that there was no defense against them. The second was that the U.S. monopoly was fleeting. It could not be sustained. The underlying technological conditions upon which the U.S. atomic program was based could be replicated elsewhere.[64] Thus the secret would not remain exclusive for long and the U.S. possession of atomic weapons would drive other states to acquire them. This problem was only exacerbated by the more basic desire of states to acquire capabilities that could also contribute to their overall economic and social development.[65]

Therefore, without a system of control, it was likely that states would attempt to develop and acquire atomic weapons to address their security needs. However, another important consideration was the incentive to acquire atomic energy production capabilities for civilian uses. Nuclear fission held the promise of a new and (potentially much more efficient) source of power generation as well as a wide array of other civilian applications. Even if states were not seeking to develop their own atomic bomb for nefarious purposes, the promise of peaceful nuclear energy would remain a powerful incentive for states to develop indigenous atomic capabilities.[66] However, because such peaceful activities cannot be differentiated from those necessary for building atomic weapons, states could never be certain of the intentions and capabilities of other states resulting in a system rife with uncertainty and mistrust. Such a situation would be highly problematic and this prospect only contributed to the argument for robust international control.[67]

The international proliferation of atomic programs—even if developed for ostensibly peaceful purposes—threatened to create a highly dangerous and unstable international system.[68] Based upon the underlying state of science and technology, atomic programs were within the reach of any state that would dedicate the necessary resources to develop them. States would have strong incentives, whether security- or development-driven to attain such capabilities. Because of the nature of atomic fission, even peaceful national programs in an unregulated world would be viewed as threatening precisely because they could also be utilized to create atomic weapons. A treaty that merely prohibited the use of atomic weapons would not provide any measure of security for the United States or other states, nor would any international agreement committing states to peaceful use of atomic energy. The distrust created by the simple existence of the capabilities of states to inflict unacceptable damage and destruction upon their neighbors in decisive first-strikes would lead to an environment of high insecurity and instability and most likely atomic war. Conventions or international agreements to prohibit atomic weapons or "ban the bomb" would be insufficient to provide security, even if supported by an international organization dedicated to policing and oversight.[69]

Without any concern for intentions, the mere existence of unregulated national atomic capabilities and the dangers created by those capabilities implied that dramatic measures were necessary to develop effective international control. In practice, this view of the potential negative ramifications of nuclear weapons also supported the position that the United States could not give up its atomic weapons until an international control organization was actually operational. Given the dangers of an anarchic world with atomic weapons, the United States could not give up its capability without a guarantee of its security. With agreement on the scope and magnitude of the problems created by atomic energy, the focus then turned to the nature of and functions of the organization necessary to be able to provide effective control. What types of powers or authority should such an organization possess? What core functions were necessary to eliminate the fears and distrust created by atomic weapons? These were the crucial questions. Robert Oppenheimer had interesting and novel answers to both.[70]

The concept of technological change was central to the discussions and ultimate approach that the Acheson-Lilienthal group pursued. "There can be no question that its dynamic changing quality is one of the dominant features of the present situation in the field of atomic energy. Advances in knowledge must be expected in a constant stream."[71] The difficulty centered on how to create an organization that could effectively manage a variety of complex issues in the context of dynamic technological change. Proposed controls or safeguards that attempted to harness or restrain technological progress or innovation in this field were inherently flawed and would prove futile. There was no putting the atomic genie back in the bottle. Only a plan that embraced the realities of atomic energy could effectively address its challenges to U.S. and international security.[72]

Most importantly, an organization that was solely relegated to the functions of oversight and inspection would not prove sufficient to the task of insuring compliance and (more importantly) deterring against potential defection. The institutional knowledge and proficiency of the staff would not be able to keep up with what was going on within national development programs if the role of the organization was relegated to these "negative" functions. For an international agency to function effectively, its staff must be, like those working within state programs, at the cutting edge of nuclear science.[73] Oppenheimer concluded that the answer might lie in an international organization that combined both the "negative" functions of inspection and control and the "positive" functions of international development. Such functions would include providing technical assistance and scientific guidance and support to individual nations as they developed their own capacities to exploit the potentially dramatic peaceful benefits of atomic energy.[74] This type of organization would be preferable to one that merely engaged in policing activities. Atomic energy held great potential for mankind, and an organization that offered to aid states in their respective quests for atomic power would be seen as attractive and desirable. An inspection agency that merely sought to maintain compliance with rules that prohibited atomic weapons would be viewed less favorably in the light of national sovereignty. Instead, an Atomic Development Authority (ADA), responsible to the United Nations, would provide a real service that states would welcome, and building upon the developmental relationships with the international community would be far better able to effectively police various national programs. "If the security organization has not only enforcement but also *development functions*, then this consideration of beneficial possibilities becomes a most weighty one."[75] This important concept was soon adopted by the group and became a central theme of the Acheson-Lilienthal report.[76]

Thus the Acheson-Lilienthal report also recommended the development of an authority staffed by highly-qualified experts from all nations that would perform the tasks of both policing against potential defection and also actively working with nations to develop peaceful civilian atomic capabilities.[77] Precisely because the pursuit of atomic energy was likely to be at the forefront of scientific inquiry, to be credible and effective, any agency would have to be staffed by those at the cutting edge of the field. Thus coupling the task of policing with the mission of aiding states in peaceful development was seen as a key to creating and maintaining a robust and relevant organization that could adapt to technological change.

As an additional safeguard against the possibility of a state seizing production or mining facilities, Acheson's group proposed that the ADA would create a "strategic balance" by distributing atomic facilities around the globe so other states would have access to atomic materials and thus deter against or (if necessary) be capable of producing their own atomic weapons. Therefore a situation where an aggressor was able to achieve its own atomic monopoly would be averted.[78] The location of these stockpiles would be decided by the United Nations. Similarly, the contributors to Acheson-Lilienthal Report understood that a charter created under the auspices of the UN-AEC would be required to develop the specific details on many of these issues, but they had accomplished a great service in setting out the major goals and key functions of an authority.[79]

The last difficult task that the committee attempted to address was the question of transition from the current unregulated system to one in which the ADA would be operational. The specific challenge of developing a schedule for various stages of progress was deemed best left to the Commission. However, the dangers inherent in the transition period worried the contributors. The dissemination of information or actual materials could provide a jumpstart to a state operating in bad faith and thus hasten the process through which that state obtained its own atomic weapon. Disclosure of information deemed essential to nuclear development should be withheld until such time as the authority was in full operation. More importantly, perhaps, there was agreement that the atomic arsenal and stockpile must remain under U.S. control until security could be guaranteed by an effective, functioning ADA.[80] Acheson and his collaborators had created a comprehensive U.S. program for the international control of atomic energy. It addressed the most pressing threat to American national security—surprise atomic attack—and created an organization that offered to resolve the problems of security and development. It also took on the challenge presented by technological change directly. While the imposition of such an organization on state sovereignty was significant, the benefits that states could obtain in return seemed to present a rational trade-off. More importantly, security and protection from atomic weapons was seemingly worth the commitment to oversight and inspection by the ADA. "It may seem idealistic. It seems time we endeavor to bring some of our expressed ideals into being. . . . It may seem too radical, too advanced, too much beyond human experience. All these terms apply with peculiar fitness to the atomic bomb."[81]

The report was presented to Secretary Byrnes on March 17 "not as a final plan, but as a place to begin, a foundation on which to build."[82] It was delivered to the President Truman on March 21, 1946, as a starting point or basis of discussion. The Acheson-Lilienthal report left important decisions such as the sequencing and specific composition of various stages toward the goal of international control unaddressed precisely because the authors believed that such decisions required technical and political investigation that would best be undertaken within the commission. Moreover, while it would remain the primary articulation of U.S. government policy guidance for the upcoming UNAEC, the question of sanctions and penalties would be revisited at the insistence of President Truman's selection to lead the U.S. delegation to the Commission, Bernard M. Baruch.

Mr. Baruch and the Question of Sanctions and the Veto

In appointing Baruch, Byrnes and Truman were counting upon his image as a senior statesman to assuage fears of key Congressional leaders about the direction of the U.S. plan in the UNAEC.[83] Baruch was well-respected across partisan lines, and could be expected to present the requisite "toughness" that skeptics in Congress believed was necessary in dealing with the Soviets in general, and in fighting for effective safeguards before any information sharing more specifically. Baruch's initial reaction to the proposed assignment focused on the challenge of Russian secrecy and security. He didn't see how the Soviets would submit to inspections if they given their strict controls on the press.[84] The most important difference between Baruch's plan and the Acheson-Lilienthal report was the lack of any sanctions or penalties associated with potential transgressions. The report did not address penalties primarily because the contributors felt that it would be the responsibility of the AEC or the larger UN organization to provide for such eventualities. More fundamentally, Acheson felt that it made little sense to talk about penalties because defection from the proposed organization would imply that cooperation have effectively broken down anyway.[85]

Ascribing penalties to such behavior when it could be expected that the entire collective organization would disintegrate, thus leaving no authority to execute them, added little value to the plan. In addition, Acheson felt focusing on penalties implied distrust of the Soviets before the cooperative offer was even made. Following from Stimson's previously articulated concerns, Acheson saw the importance of presenting the plan in good faith, and that the Soviets perceive it as such, if there was any chance of real progress on the issue.[86] Thus penalties or sanctions were a nuisance that provided no value to the plan and could only poison the atmosphere of negotiations.[87]

To Baruch and his associates, sanctions were essential.[88] Clearly starting from a different premise than Acheson, it was argued strongly that the UN-AEC and the UN itself must be imbued with the powers necessary to punish bad behavior. As it was written, the Acheson-Lilienthal report provided no protections. It only provided for warning, and the impact of this warning in such a crucial matter was insufficient. If the United States was to support the creation of an international institution that was truly capable of providing for its security and the security of the world, that institution must be empowered with the capabilities necessary to deter, and if necessary, defend against aggression. Merely providing warning against aggression was not nearly enough, given the stakes involved.[89] Such punishment should be swift and certain. The automatic nature of the penalty was driven by Baruch's own views of the failings of the League of Nations and he was determined that the failures of the past would not be repeated by the UN.[90]

As he later wrote, "If I had learned anything out of my experience in international affairs, it was that world peace is impossible without the force to sustain it. Consequently, I insisted that any plan for the control of atomic energy contain provision for sanctions against those who violated the rules."[91] Of course, Acheson would argue that Baruch had learned the "wrong lesson." Baruch's response to the failing of past cooperation was to attempt to create a more powerful, more robust international organization, capable of deterring or compelling good conduct among its members. Acheson, perhaps more realistically, focused on the need to make the proposed organization attractive to individual states. The ADA would succeed or fail on the basis of how states perceived the potential benefits and protections provided by engaging in cooperation. It was important that cooperation be presented to the international community (and the Soviets in particular) as something that would further their national interests. Moreover, the United States must avoid the creation of a perception that the UNAEC and the ADA were merely tools of American interests. This was the fundamental difference in the approaches of the Acheson and Baruch groups.[92]

After consideration of the differing views, President Truman agreed with Baruch and Byrnes.[93] Given the importance of the atomic issue and the ongoing difficulties in negotiating with the Soviets, Truman favored a stronger approach to the upcoming negotiations. He was persuaded that an effective agency should have the powers to punish potential transgressors. Baruch was to have the authority to advocate the position of the United States as he saw appropriate, though Truman made clear that he expected to be consulted on any major issues that might arise.[94] Beyond the penalty and sanctions issues, the major components for a plan of international control remained remarkably consistent throughout the policy process. If anything, the tone of the Baruch Plan may have differed from the initial Acheson-Lilienthal report, but this is hardly surprising. The Acheson Committee and the board of consultants deliberately insulated their analysis from the political constraints of the emerging Cold War.[95]

NEGOTIATIONS ON THE INTERNATIONAL CONTROL OF ATOMIC ENERGY

On June 14, 1946, first session of the UNAEC convened in the gymnasium of Hunter College in New York City. Represented were delegates and staff from the permanent five Security Council members (the United States, Great Britain, France, the Soviet Union, and China) as well as Canada, Australia, Brazil, Egypt, France, Mexico, the Netherlands, and Poland. Baruch, representing the United States, was the first delegate to speak after UN Secretary-General Trygve Lie officially opened the proceedings. Reading from text drafted by Herbert Swope, Baruch presented the U.S. position with a tone of urgency and gravity commensurate with the challenge at hand: “We are here to make a choice between the quick and the dead. That is our business.”[96]

In presenting the U.S. proposal, which embodied the underlying assumptions and primary goals of the Acheson-Lilienthal report, Baruch clearly underscored the importance of the administration’s views on the necessity of sanctions and penalties which had created disagreement within the U.S. policy community.[97] Moving on from this dramatic opening, Baruch presented the details of the U.S. plan, including the specific fourteen-points that had been distilled from the Acheson-Lilienthal report. His presentation also included a brief discussion of the larger objectives that would be achieved after the successful construction of the Atomic Development Authority:[98] Immediately following this section, Mr. Baruch provided an explicit list of potential violations that would necessitate automatic sanction by the United Nations.[99] His discussion of the importance of penalties and sanctions left no ambiguity about where the United States stood on the question of the veto in the UN Security Council:

> I want to make very plain that I am concerned here with the veto power only as it affects this particular problem. There must be no veto to protect those who violate their solemn agreements not to develop or use atomic energy for destructive purposes. The bomb does not wait upon debate. To delay may be to die.[100]

While the tone and focus of Baruch's presentation received criticism both within and outside of the Truman administration, it clearly reflected the broad consensus of thinking across the U.S. civilian and military leadership.[101] The nature of atomic weapons was such that nothing less than completely effective international control was acceptable.

The United States was willing to support the development of an ADA that would not only protect against potential violations, but also facilitate the dissemination of scientific and technical information, and aid in the creation and development of national atomic programs. However, in return for this proactive and generous offer by the United States to the rest of the world, American security must be guaranteed. Only if a comprehensive program explicitly premised upon a voluntary surrender of national sovereignty was adopted, could the problem of atomic war be solved.[102] The United States was willing to follow through if others (most importantly the Soviets) reciprocated its offer, but it was unwilling to surrender its own atomic arsenal, the central component of its national security to do so. The Baruch Plan and the overall U.S. support for the United Nations and the AEC were perceived to be sufficient gestures of good faith and a commitment to cooperation. Despite Baruch's statement that the plan was "an outline for present discussion," in reality it amounted to an ultimatum, a "take it or leave it offer." But for those responsible for the security of the United States, it was the only logical approach. The inherent dangers created by atomic weapons precluded half-measures or partial agreements that would fail to reach the perceived threshold for cooperative security deemed essential for the safety of the United States. The only real question was whether the comprehensive nature of the plan and the potential benefits to other states contained within it would prove to be enough of a signal of good faith, and not merely cheap talk or diplomatic gamesmanship.

The official responses to the Baruch Plan were made during the June 19 session of the UNAEC. As expected, it received generally strong support from the representatives of Canada, Great Britain, China, Brazil, and Mexico.[103] The representative from the Soviet Union, Andrei Gromyko responded next. After briefly discussing the importance of the task at hand and acknowledging the revolutionary importance of atomic energy for both international security and peaceful civilian development, Gromyko presented the Soviet view on how to proceed:

> As one of the primary measures for the fulfillment of the resolution of the General Assembly of 25 January 1946, the Soviet Delegation proposes that consideration be given to the question of concluding an international convention prohibiting the use of atomic energy for the purpose of mass destruction. The object of such a convention should be the prohibition of the production and employment of atomic weapons, the destruction of all existing stocks of atomic weapons and the condemnation of all activities undertaken in violation of this convention.[104]

Thus, in his initial remarks, Gromyko presented the very position that United States policy had viewed as fundamentally unacceptable. Taking the opposing view to the stated U.S. position, Gromyko's demand for a prohibition of atomic weapons as a necessary first step for negotiations to develop a control organization was completely incompatible with the U.S. proposal. Seeing the U.S. bomb as the primary threat to Soviet security, it was not surprising the Gromyko would make "banning the bomb" a requisite for Soviet cooperation in the development of an international organization.[105]

However, as discussed above, this was completely antithetical to the American view that its atomic capabilities could be addressed only after an effective control system was operational. While other delegates and observers may have seen some possibility for progress if these "sequencing" issues could be resolved, in reality neither the United States nor the Soviets were in a position to compromise. The issue was indivisible.[106]

Gromyko's attack on the Baruch proposal to eliminate the veto was unsurprising. From the Soviet point of view, the Security Council veto was the only guarantee that the United Nations could not be transformed into a front for the diplomatic advantage of the capitalist powers. If it were to relinquish the veto, the Soviet Union would leave itself completely vulnerable to the decisions of the United States. It would, in essence, subordinate its national interests and security to the good graces of Washington. Given the wide range of issues remaining to be settled between the two superpowers, the Soviets feared that disagreement over the postwar settlements could be used as a premise to punish the Soviet Union through the mechanism of the United

Nations. While the surrender of the veto would ostensibly be applied only to matters within the scope of the UNAEC, there was no guarantee that the U.S. would not seek to link Soviet behavior in other areas to that forum. For these reasons, elimination or even moderation of the Security Council veto was a "non-starter." Gromyko asserted as much in private meetings with Baruch.[107]

Despite the vast differences between the U.S. and Soviet proposals, both sides attempted to maintain a working relationship, providing the perception that progress could be made.[108] It was decided that the larger political issues associated with cooperative international control of atomic energy would be addressed within a working committee that became known merely as "Committee Two" because of wrangling about the proper title and scope of the committee's responsibilities. Meanwhile, a Scientific and Technical Committee would concurrently discuss and report on those aspects of the atomic energy problem.[109]

The Science and Technical Committee met eighteen times between August 1 and October 7. It ultimately submitted a report on September 20.[110] The document was almost identical to the Acheson-Lilienthal report. With the work of the Science and Technical Committee concluded, the focus moved back to the political issues. At this point, Baruch began to concede that agreement was not likely to be forthcoming and sought guidance from President Truman and Secretary Byrnes on how to proceed. The Soviets wanted a ban on the atomic bomb before proceeding with negotiations toward an ADA, while the U.S. position clearly demanded a functioning organization as a requisite to any limitations on its atomic arsenal.[111] The positions would not change. Nonetheless, Committee Two met nineteen times between October 15 and December 13.[112] The Soviet representatives consistently argued that they could not understand why the United States was unwilling to stop producing atomic weapons.[113] They also viewed the U.S. proposal as "too ambitious" and antithetical to concepts of national sovereignty.[114] Conversely, the American delegation argued that only Soviet acceptance of such a proposal could make bomb production unnecessary.

Negotiations Hit an Impasse

At this point it was abundantly clear that the neither side was going to move off of its positions and that negotiations were deadlocked. A direct approach to Stalin was briefly considered and rejected.[115] Baruch felt that the window of opportunity for the United States to gain at least a nominal victory in a vote on the report was closing.[116] After consulting with President Truman, Secretary Byrnes supported Baruch's position. On November 5, he instructed Baruch to proceed with a vote before the end of the year to avoid a further delay that may come with changes in the Commission after the New Year.[117] Thus, at its November 13 meeting, the AEC decided to submit a report to the

Security Council by the end of the year. On December 5, Baruch presented the formal plan for control to be included in the commission's report. A "First Report on Safeguards" had been produced with supporting memoranda and technical papers provided by the U.S. delegation. The report was formally discussed on December 18, 19, and 26.[118] On December 17, Gromyko attacked the veto concept and the fact that the United States had not agreed to prohibit the bomb. He asked for additional time for further discussion and negotiation. From December 20 on, the Soviets effectively abstained from the proceedings.[119] On December 30, the first AEC report passed 10-0 with the Soviet and Poland abstaining.[120]

The UNAEC continued to explore various options throughout the next two years, but no progress was made. Mr. Baruch formally submitted his resignation to President Truman on January 4, 1947. He and his staff had guided the U.S. proposal through an important first phase, but he believed that with the victory in the commission, it was time to depart.[121] It is difficult to see what progress Baruch would have believed possible upon his departure.

Whether Baruch's insistence on including sanctions and the suspension of the veto undermined the cooperative spirit of the negotiations, the differences between his presentation and the Acheson-Lilienthal report were marginal at best. His representation of the U.S. position was true to the underlying agreement within the Truman administration that the only effective system of international control would be one that was complete, transparent, and essentially absolute. The system had to be foolproof or the United States could not give up its atomic weapons. The Soviets were unwilling to agree to any system before the United States relinquished its arsenal. This was the crux of the problem. Tone ultimately mattered little. To his last official word, Baruch tenaciously maintained his commitment to what was effectively the consensus U.S. position. Complete and total international control or nothing at all. It was that simple. In the absence of that development and Soviet capitulation on the issue, the United States should maintain and expand its monopoly to the extent possible.[122]

The Soviet View of Negotiations

Given the premises of Soviet Policy, it was highly unlikely that Stalin or his colleagues believed that international control would be established. They did not expect help from the United States in building the bomb; nor did they expect the United States to give up its monopoly. On the contrary, they expected the United States to try to hold on to its monopoly for as long as possible, and to use it to pressure the Soviet Union.[123] Therefore Gromyko's insistence on the need for a convention to prohibit atomic weapons, to "ban the bomb," is not surprising. It reflected the security interests and diplomatic preferences of the Soviet Union and its leader Josef Stalin. Given their position of relative inferiority vis-à-vis the United States in terms of atomic research and development, it seems that only such a gesture as the voluntary unilateral atomic disarmament of the American adversary could provide an adequate signal of good faith. In reality, even such a dramatic and irrational gesture as that may not have sufficed. As MET would predict, Stalin was unlikely to be satisfied by any political agreement other than one that would contribute to, or at least allow for, the development of a Soviet atomic bomb. Acquisition of the atomic bomb was a national priority, and the only true means to reestablish a balance between the Russia and America. Locking the USSR into a position of inferiority in the field of atomic energy, particularly given the decisive military applications of that technology, was inconceivable to Stalin. It was irrational to accept the U.S. proposal and thus depend on Washington's good faith.

Moreover, the perceived American attack on the Soviet veto power in the UN Security Council seemed to signal lack of faith and possibly malign intentions. It reaffirmed Soviet fears of capitalist encirclement. However, in reality, the Soviet position on atomic control had been determined long before Baruch's proposal in the UNAEC, and well in advance of the London and Moscow meetings that had created the structure to facilitate the cooperative goal of international control of atomic energy. It is more likely that the Soviet policy on atomic weapons had been set between July and August 1946; the period between Truman's announcement of the U.S. bomb to Stalin at Potsdam and the bombing of Hiroshima. Two important factors driving the Soviet approach to atomic weapons precluded cooperation: Stalin viewed the weapon as decisive and thus central to Soviet security, and the underlying technological conditions provided the Soviets with an opportunity to obtain it.[124]

The diplomatic strategy for managing relations with the United States in the wake of their successful test seems to involve two distinct and seemingly countervailing approaches. The Soviets cooperated with their erstwhile allies in the creation of the organization that would undertake the challenge of international control of atomic energy, while at the same time energetically and vociferously articulating positions that protected Soviet interests.[125] Thus the Soviets exhibited a willingness to engage the United States in negotiations to cooperatively resolve the challenges of controlling atomic energy, but they were unwilling to make concessions that would leave them relegated to an inferior position in terms of security and technological development. The Soviets could offer a convention prohibiting atomic weapons, precisely because such a convention would have no real power to restrain the Soviet development program. A system of international control, however, threatened to do just that, and was thus unacceptable. The threat to Soviet security presented by the U.S. atomic bomb could only be rectified by a

Soviet bomb, and the confirmation that the Soviet Union possessed the scientific and technological capacity to remain on an equal footing with the United States. This does not necessarily mean that Stalin sought the bomb for nefarious purposes, or to further the Soviet Union's ability to launch an aggressive, revisionist campaign. It is only to argue that until such time as the Soviet Union had successfully developed its own atomic bomb, a new decisive weapon which could change the nature of warfare, Stalin believed that the Soviet Union was vulnerable. Perhaps, at that time, the question of international control could be revisited, but until then any cooperative initiative that sought to impose a limitation on the Soviet Union's capacity to successfully develop an atomic capability was unacceptable. As Gromyko makes clear:

> The "Baruch Plan" (as the press dubbed it), which he presented but which was really the Pentagon's plan, boiled down to making sure that the USA retained its monopoly on nuclear weapons. The intention was that the USSR and the rest of the world should to a significant extent place their security in Washington's hands. The USSR found this unacceptable.[126]

By the time of the negotiations in the UNAEC, the Soviet atomic program was already running at a high clip.[127] While Hiroshima and Nagasaki had shocked Stalin (and much of the rest of the world), the use of atomic weapons provided two distinct rationales for engaging in an intensive crash program to create a Soviet bomb. The first, and more obvious, is that the detonation of a U.S. bomb provided an incentive for Stalin and the Soviet scientists to obtain a weapon quickly to "catch up" to the United States and thus close a dangerous window of vulnerability. A second but equally important incentive that arose directly from the success of the U.S. bomb was the affirmation that the development and practical utilization of such a weapon was indeed possible. The Soviet program, initiated in 1943 in response to intelligence that Germany and the United States were both working on the development of an atomic bomb, obviously suffered as a result of the nature of the war.[128] It was difficult to argue for the allocation of scarce resources for a theoretical weapon while German soldiers remained on Soviet soil. Moreover, the quality of the information that Stalin and his intelligence chief Beria received was inconclusive insofar as there was little hard evidence that these nascent atomic programs would actually create a "usable" weapon in a relevant time frame. There were other, far more pressing concerns. After Truman's casual acknowledgment to Stalin that the United States possessed an atomic weapon, the Soviet leader immediately reacted by implementing a full-scale program, tightly controlled under Beria to provide the necessary support and to facilitate access to resources.

From this time on, the clock was ticking and time was moving against the Soviet Union. Stalin sought to achieve his various objectives before the

United States could end the war with an atomic strike on Japan. Time ran out, but Stalin was unwilling to resign himself to this new situation. Instead he pushed harder to develop a Soviet bomb precisely because he viewed it as a decisive weapon, without which the Soviet Union would be vulnerable. However, with the explosion of the U.S. atomic bomb, Stalin and the scientific and bureaucratic actors whom he had entrusted with the Soviet bomb program now knew for certain that such a weapon was attainable. The U.S. bomb confirmed that such a goal could be accomplished. Moreover, with the significant benefits of espionage, the Soviets were able to understand and build upon what the United States had accomplished.[129] The Soviets now knew that they were "on the curve" and though they might be behind the United States, they could catch up with the necessary commitment of resources.

Technological change had placed the United States in a position of advantage, and while most acknowledged that the advantage would ultimately matter little in the face of an arms race that could destroy the world, America only had a limited range of options. Unless the Soviets were willing to agree to an intrusive and wholly transparent system of international control and monitoring, the United States preferred to maintain its monopoly and continue to develop atomic weapons.

CONCLUSIONS

In retrospect, the revolutionary redefinition of state sovereignty required for the successful implementation of the Baruch Plan seems to preclude its adoption in an anarchic realm inhabited by nation states. However, the conclusion drawn by a broad consensus of the U.S. political-military policy community was precisely that the revolutionary nature of atomic weaponry demanded such a redefinition of traditional state sovereignty in order to maintain peace and avoid nuclear disaster. Moreover, the documentary evidence clearly reflects the fact that the key policymakers in the United States were willing to give up the U.S. atomic monopoly, but only if a completely effective international control regime could be developed. The Baruch Plan was not merely an attempt to strike a first, meaningful blow in what would become an increasingly vitriolic propaganda battle between the United States and the Soviet Union.

The atomic bomb, holding the potential to be perhaps the most decisive of all weapons, instilled fears even in those who possessed it first, without immediate challenge. The specific fear of the development and deployment of atomic weapons by another state, particularly the Soviet Union, was so disconcerting to the Truman administration that it sought to establish effective international control to address the longer-term threat created by atomic weapons. As the evidence shows, very soon after the initial successful test of atomic fission, key policymakers in the U.S. leadership saw the potential vulnerabilities to the United States that would be created in a world where other states possessed atomic arsenals. Bolstering this concern was the consensus opinion within policymaking circles that the scientific and technical knowledge would not and could not remain exclusive to the United States.

The curve had shifted, but other states would soon be capable of catching up to the United States. Moreover, the advantage that the United States possessed in atomic weaponry would provide little additional security in the face of an enemy with atomic weapons. The one solution to this daunting challenge seemed to be the development of an effective international control regime that would prohibit atomic weapons and prevent a dangerous and potentially destabilizing arms race. Though cooperation had failed to prevent the last war, and despite the difficulties that the United States and Soviets were experiencing on other diplomatic issues, the United States attempted to incorporate their former allies into an institutional arrangement that would have aided states in the development of the civilian uses of atomic energy and remove atomic weapons from world politics. The U.S. atomic monopoly provided a limited and temporally bounded advantage and the Truman administration was willing to forego it should cooperation prove possible, but it was equally clear that without the guarantees inherent in an effective system of control and inspection, America would not give up its atomic arsenal.

While the onset of the Cold War may seem to over-determine the failure of the UNAEC, the documentary evidence and process-tracing implies that MET accurately explains the political dynamics involved. Given the centrality of atomic weapons to the Cold War system, the deliberate attempt to delink the question of international control from other diplomatic issues and the importance placed upon it by the Truman administration, MET provides a useful lens for understanding the case. The year 1946 was a time of high uncertainty, but the negotiations to develop international control of atomic energy were not driven by these political concerns. If anything, the failure of the Baruch Plan and the AEC was a contributing factor, rather than a result of the Cold War. The failure of the Baruch Plan was another instance of unsuccessful cooperation that each side integrated into its changing perception of the other, leading to the open hostility that would emerge in 1947 and 1948. However, this failure was due primarily to the inability of either state to essentially surrender their position and accept the other's as their own. The decisiveness of atomic weapons were such that the United States could not accept any plan that did not implement controls first and the Soviet could not accept a plan that restrained their ability to at least develop atomic weapons. These were mutually exclusive and indivisible bargaining positions.

The history of the UNAEC and the Baruch Plan provides strong support for MET's first general proposition which would predict failure in the presence of decisive weapons while also supporting the first specific hypothesis focusing on the motivations of a leading state to broach the possibility of qualitative arms control under relatively favorable political-military conditions. The United States was so fearful of the atomic bomb that it sought to develop a foolproof system of international control to prevent its proliferation. However, because of inherent difficulties in confidently detecting cheating or diversion—even with an expansive oversight organization—cooperation proved impossible. The Soviet experience similarly supports the specific hypotheses of laggard state views on decisive weapons. Offensive realist hypotheses may again explain the failure, but certainly cannot explain the proposed cooperation or the motives or behavior of the United States and the military and political leaders of the Truman administration. Similarly, defensive realism may predict failure but would face a challenge in effectively telling the story of that failure. The role of the United Nations and the Atomic Energy Commission would seemingly provide a robust role for institutions, but the inability of the proposed organization to allay the fears of members and provide for their security reinforces the basic challenge for institutions in the security realm and the limited relevance of the approach for understanding security cooperation.

The Soviet counter-proposal demanded the reverse: an abolition of atomic weapons before cooperation could be developed. While it seems that Stalin never intended to forego the opportunity to develop a Soviet atomic bomb, and that the security of developing an indigenous weapon was a prerequisite to any international cooperation, this only reinforces the MET's central hypotheses. A state facing a breakthrough by a potential rival will vigorously pursue that technological application unless it is fundamentally incapable of doing so. Thus the case strongly conforms to the predictions of MET and its focus on the decisiveness of a given military asset. There is little doubt that leaders on both sides viewed the atomic bomb as a revolutionary weapon. The only cooperation that could possibly address the threat to security presented by such a weapon would necessarily have to incorporate a revolutionary redefinition of sovereignty and national authority. In an anarchic international system, defined by uncertainty and mistrust, this was simply a leap too far.

NOTES

1. Rhodes, *The Making of the Atomic Bomb*, Martin J. Sherwin, "The Atomic Bomb and the Origins of the Cold War: U.S. Atomic-Energy Policy and Diplomacy, 1941–45," *American Historical Review* 78, no. 4 (1973).

2. Alperovitz, *Atomic Diplomacy*, 220–38, Martin J. Sherwin, "The Atomic Bomb and the Origins of the Cold War," in *Origins of the Cold War: An International History*, eds. Melvyn P. Leffler and David S. Painter (London: Routledge, 1994), Sherwin, "The Atomic Bomb and the Origins of the Cold War"; Sherwin, *A World Destroyed.*

3. See “Minutes of a Meeting of the Combined Policy Committee,” *FRUS, 1945* I, 86–89; Hewlett, *The New World*, 459–69.
4. “Draft Proposals on Atomic Energy for Submission to Soviet Government,” December 10, 1945 in *FRUS, 1945, II*, 92–96.
5. Hewlett, *The New World*, 473–77; Lieberman, *The Scorpion and the Tarantula*, 204–24.
6. Herken, *The Winning Weapon*, 61.
7. See “The Ambassador in the Soviet Union (Harriman) to the Secretary of State,” November 27, 1945, in *Foreign Relations of the United States, 1945, Europe*, vol. 5 (Washington, DC: U.S. Department of State), 922–24.
8. Byrnes, *Speaking Frankly*, 110–22; Hewlett, *The New World*, 475–77.
9. Acheson, *Present at the Creation*, 136; *FRUS, 1945, II*, 609–10; David McCullough, *Truman* (New York: Touchstone, 1992), 476–80; Truman, *Memoirs of Harry S. Truman*, 547–52.
10. *International Control of Atomic Energy: Growth of a Policy* (Washington, DC: U.S. Department of State, 1946), 29–33.
11. Hewlett, *The New World*, 580–82.
12. Kennan’s “Long Telegram” can be found in *Foreign Relations of the United States, 1946, Eastern Europe, the Soviet Union*, vol. VI (Washington, DC: U.S. Department of State), 696–709. The Clifford-Elsey Report (September 1946) is reprinted in Arthur Krock, *Memoirs: Sixty Years on the Firing Line* (New York: Funk & Wagnalls, 1968), 419–82.
13. Robert Gilpin, *American Scientists and Nuclear Weapons Policy* (Princeton, NJ: Princeton University Press, 1962), 40–47; Jungk, *Brighter Than a Thousand Suns*, 173–74. The memorandum is reprinted in an appendix, 343–47.
14. Jungk, *Brighter Than a Thousand Suns*, 178–80.
15. Ibid., 183–85.
16. Ibid., 351.
17. Herken, *The Winning Weapon*, 11–22; Hewlett, *The New World*, 347–407.
18. Jungk, *Brighter Than a Thousand Suns*, 349.
19. Ibid., 358.
20. Hewlett, *The New World*, 415–21.
21. See “Memorandum from Secretary of War (Stimson) to President Truman,” September 11, 1945, in *Foreign Relations of the United States, 1945, General Political and Economic Matters*, vol. II (Washington, DC: U.S. Department of State), 43.
22. Ibid., 44.
23. Ibid.
24. Henry L. Stimson and McGeorge Bundy, *On Active Service in Peace and War* (New York: Harper, 1948).
25. Harry S. Truman, *Memoirs of Harry S. Truman: 1945 Year of Decisions*, vol. 1 (New York: Doubleday, 1955), 416.
26. James F. Byrnes, *Speaking Frankly* (New York: Harper & Brothers, 1947), 263.
27. Holloway, *Stalin and the Bomb*, 116–17; Zubok and Pleshakov, *Inside the Kremlin’s Cold War*, 39–46.
28. See Stimson to Truman, September 11, 1945, *FRUS, 1945, II*, 41.
29. Ibid., 42. Italics added.
30. On Acheson, see, for example, Dean Acheson, *Present at the Creation: My Years in the State Department* (New York: W. W. Norton, 1987); Robert L. Beisner, *Dean Acheson: A Life in the Cold War* (Oxford: Oxford University Press, 2006); David S. McLellan, *Dean Acheson: The State Department Years* (New York: Dodd, Mead, 1976).
31. See “Memorandum from Undersecretary of State (Acheson) to President Truman,” September 25, 1945 *FRUS, 1945, II*, 48.
32. Ibid.
33. Ibid., 48–49.
34. Ibid.
35. Ibid.

36. See "Minutes of a Meeting of the Secretaries of State, War, and Navy," October 10, 1945 in *FRUS, 1945, II*, 56. The delay would ultimately foreclose Stimson's initial plan of approaching the Soviets directly.
37. Hewlett, *The New World*, 423–28.
38. Ibid., 532.
39. Ibid., 532–33.
40. See "Memorandum by General Leslie Groves: 'Our Army of the Future: As Influenced by Atomic Weapons,'" January 2, 1946, in Hewlett, *The New World*, 1197–203.
41. Ibid., 1198.
42. Ibid., 1199.
43. Ibid.
44. Ibid., 1198.
45. Ibid., 1201.
46. Ibid., 1198.
47. On the incentives toward preventive war with the Soviets see for example: Richard K. Betts, "A Nuclear Golden Age? The Balance before Parity," *International Security* 11, no. 3 (1986–1987); Russell D. Buhite and W. Christpoher Hamel, "War for Peace: The Question of an American Preventive War against the Soviet Union, 1945–1955," *Diplomatic History* 14, no. 3 (1990); David Alan Rosenberg, "The Origins of Overkill: Nuclear Weapons and American Strategy, 1945–1960," *International Security* 7, no. 4 (1983); Marc Trachtenberg, "Strategic Thought in America, 1952–1966," *Political Science Quarterly* 104, no. 2 (1989).
48. Groves memorandum, *FRUS, 1946,* I, 1198.
49. Ibid.
50. Ibid., 1197.
51. See "Memorandum by the Joint Chiefs of Staff to the State-War-Navy Coordinating Committee," January 23, 1946, in *FRUS, 1946, I*, 739.
52. See "Foreign Policy of the United States: Atomic Energy," in *Foreign Relations of the United States, 1946, General: The United Nations*, vol. 1 (Washington, DC: U.S. Department of State), 712–16.
53. Truman, *Memoirs of Harry S. Truman: 1945 Year of Decisions*, 416.
54. Byrnes, *Speaking Frankly*, 263.
55. Georgii Konstantinovich Zhukov, *The Memoirs of Marshal Zhukov* (New York: Delacorte Press, 1971), 674.
56. Gromyko, *Memoirs*, 109.
57. Ibid., 110.
58. Holloway, *Stalin and the Bomb*, 147–49.
59. Zubok and Pleshakov, *Inside the Kremlin's Cold War*, 43–44.
60. Ibid., 149–52.
61. I. V. Kurchatov, "Notes on the Discussion between I. V. Kurchatov, Lead Scientist for the Soviet Nuclear Effort, and Stalin, 01/25/1946," (Cold War International History Project, 2007).
62. Ibid.
63. See "Minutes of the Meeting of the Secretaries of State, War, and Navy," January 24, 1946, in *FRUS, 1946,* I, 736–38; Hewlett, *The New World*, 533–34.
64. Ibid., 538.
65. "A Report on the International Control of Atomic Energy," (Washington, DC: U.S. Department of State, 1946), 2.
66. Hewlett, *The New World*, 542.
67. "A Report on the International Control of Atomic Energy," 4.
68. Hewlett, *The New World*, 542–45; "A Report on the International Control of Atomic Energy," 4–5, 33.
69. "A Report on the International Control of Atomic Energy," 4–5.
70. See "Memorandum by Dr. J. Robert Oppenheimer," *FRUS, 1946,* I, 749–54.
71. "A Report on the International Control of Atomic Energy," 15.
72. See Oppenheimer Memorandum in *FRUS, 1946,* I, 751.
73. Ibid.

74. Ibid., 749–54.
75. "A Report on the International Control of Atomic Energy," 18.
76. Ibid., 17.
77. Ibid., 45.
78. Ibid., 47.
79. Ibid., 33.
80. Ibid., 51–56.
81. Ibid., 31.
82. See "The Secretary of State's Committee on Atomic Energy to the Secretary of State," March 17, 1946, in *FRUS, 1946,* I, 761–64.
83. Hewlett, *The New World*, 554–58; Lieberman, *The Scorpion and the Tarantula*, 260–65; Truman, *Memoirs of Harry S. Truman*, 7–8.
84. See "Mr. Bernard M. Baruch to President Truman," *FRUS, 1946,* I, 767–68.
85. See "Memorandum of Conversation by Mr. John Hancock of the United States Delegation to the Atomic Energy Commission," May 29, 1946, in *FRUS, 1946,* I, 804.
86. Acheson, *Present at the Creation: My Years in the State Department*, 155–56.
87. See "Memorandum by the Under Secretary of State (Acheson) to the Secretary of State," June 6, 1946, in *FRUS, 1946,* I, 836.
88. Hewlett, *The New World*, 567–74.
89. See Hancock Memorandum, *FRUS, 1946,* I, 804.
90. Ibid., 813.
91. Bernard M. Baruch, *Baruch: The Public Years* (New York: Holt, Rinehart and Winston, 1960), 366.
92. See "Memorandum by the United States Representative on the Atomic Energy Commission (Baruch) to President Truman," June 6, 1946, in *FRUS, 1946,* I, 839.
93. Hewlett, *The New World*, 574; Truman, *Memoirs of Harry S. Truman*, 10–11.
94. See "Memorandum by President Truman to the United States Representative on the Atomic Energy Commission (Baruch)" June 7, 1946, in *FRUS, 1946, I*, 846.
95. *Documents on Disarmament, 1945–1959* (Washington, DC: U.S. Department of State, 1960), 10–11.
96. *Documents on Disarmament, 1945–1959*, 7–8.
97. Ibid., 8.
98. Ibid., 11.
99. Ibid., 11–12.
100. Ibid., 12.
101. Byrnes, *Speaking Frankly*, 270–74; Herken, *The Winning Weapon*, 175–76; Hewlett, *The New World*, 578–79; Truman, *Memoirs of Harry S. Truman*.
102. The plan received significant support from the scientific community. Gilpin, *American Scientists and Nuclear Weapons Policy*, 52–60.
103. Hewlett, *The New World*, 582–83.
104. *Documents on Disarmament, 1945–1959*, 18.
105. Ibid.
106. Andrei Gromyko, *Memoirs* (New York: Doubleday, 1989), 138.
107. Gromyko, *Memoirs*, 140.
108. The negotiations are covered in detail in Herken, *The Winning Weapon*, 174–91; Hewlett, *The New World*, 585–619; Lieberman, *The Scorpion and the Tarantula*, 315–86.
109. Baruch to Truman, *FRUS, 1946,* I, 921.
110. Hewlett, *The New World*, 588–94.
111. Baruch to Truman, *FRUS, 1946,* I, 924–25.
112. Hewlett, *The New World*, 586–96.
113. See "Memorandum by Dr. Richard C. Tolman to Mr. John M. Hancock," October 4, 1946, in *FRUS, 1946,* I, 951–52.
114. See "Memorandum by Mr. Franklin A. Lindsay to the United States Representative on the United Nations Atomic Energy Commission," October 21, 1946, in *FRUS, 1946,* I, 955–60.
115. Kennan broached the idea and offered to facilitate a meeting. See "Memorandum of Conversation, by Mr. Franklin A. Lindsay," November 12, 1946, in *FRUS, 1946,* I, 1011.

116. See "The United States Representative on the Atomic Energy Commission (Baruch) to the Secretary of State," November 4, 1946, in *FRUS, 1946,* I, 991.
117. See "the Secretary of State to the United States Representative on the Atomic Energy Commission (Baruch)," November 5, 1946, in *FRUS, 1946,* I, 995.
118. Hewlett, *The New World*, 611–16.
119. Ibid., 613–17.
120. Ibid., 617–19.
121. See "The United States Representative on the United Nations Atomic Energy Commission (Baruch) to President Truman," in *Foreign Relations of the United States, 1947, General: The United Nations*, vol. 1 (Washington, DC: U.S. Department of State), 332–33.
122. Ibid., 336.
123. Holloway, *Stalin and the Bomb*, 162.
124. Ibid., 116–18; Zubok and Pleshakov, *Inside the Kremlin's Cold War*, 41–44.
125. Holloway, *Stalin and the Bomb*, 169.
126. Gromyko, *Memoirs*, 138.
127. Holloway, *Stalin and the Bomb*, 162.
128. Ibid., 90–95.
129. Ibid., 103–8.

Chapter 5

The SALT I Accords

The Strategic Arms Limitation Talks (SALT) process spanned over fifteen years and resulted in two arms control treaties, SALT I and SALT II, although the latter was never ratified by the U.S. Senate. The process was characterized by arduous negotiations between the United States and the Soviet Union over the qualities and quantities of their respective nuclear (strategic) arsenals. Initiated by the Johnson administration, the talks focused primarily on the limitation or proscription of Antiballistic Missile (ABM) systems, which had proven to be potentially destabilizing (if either side actually achieved a breakthrough) but unattainable in any relevant time frame. President Richard M. Nixon signed the SALT I accords with Soviet Premier Leonid Brezhnev in Moscow in May 1972.[1] The centerpiece of the SALT I accords was the ABM treaty, which placed strict limitations on missile defense programs. In addition, the superpowers concluded an Interim Agreement on Strategic Offensive Arms, which formalized agreed numerical ceilings on offensive ballistic missile systems. Nixon presented the accords as a major diplomatic achievement and it did seem to provide two clear security benefits.

First, in first limiting national missile defense systems, the ABM Treaty allowed both superpowers to relax efforts to achieve a breakthrough in strategic defenses and thus avoid a new, potentially destabilizing round of the arms race. Second, the limitations on offensive weapons seemed to formally acknowledge the importance of a relatively stable strategic balance, where both superpowers would have the ability to maintain a guaranteed second-strike and thus an assured destruction capability. The major perceived failure of SALT I was the inability to agree to limitations on one key innovative weapon system, the Multiple Independently-targeted Reentry Vehicle (MIRV), which soon emerged as the spark for what could create a new, intense round of the arms race.

This chapter will investigate the SALT I accords, which included the ABM Treaty and Interim Agreement. The case provides several useful tests for MET. Despite representing a potentially decisive new application, significant time and resources had been devoted to developing viable missile defense systems. Both Washington and Moscow were relatively confident that strategic defenses would not be achieved in any relevant time frame, but so long as the other was seen to be seriously working on the problem, fears of a breakthrough prevented either from abandoning their efforts. Thus, expectations were that missile defenses would not be decisive in any relevant time frame, and also that limitations (or prohibitions) on the components of these systems could be confidently monitored by unilateral national technical means. In doing so, a potentially destabilizing new round of the arms race could be avoided through cooperation. Similarly, on the offensive side of the ledger, delivery vehicles—ICBMs and SLBMs—had achieved a level of technological maturity that leaders expected while they would continue to be central to their nations' respective capabilities, limitations could enhance stability and contribute to mutual deterrence, thereby decreasing the probability of a conflict. Again, leaders on both sides were confident that limitations could also be effectively verified by national technical means, preventing either side from achieving a quantitative breakout that could dramatically shift the balance.

Conversely, the story of MIRV strongly affirms the competitive hypotheses of MET. Viewing MIRV as a decisive new application in the superpower arms race, the United States undertook a concerted effort to develop and deploy these warheads with little or no interest in seeking cooperative means to limit or prohibit them. MIRV provided a significant hedge against the ongoing Soviet buildup as well as a guarantee against a potential Soviet breakthrough in missile defenses. Similarly, the Soviet Union, seeing decisive capabilities of MIRV for its own strategic arsenal, had no interest in locking itself into a position of technological inferiority in the face of an extensive U.S. deployment. Only when U.S. leaders realized that extensive deployments of Soviet MIRV could threaten the land-based U.S. ICBM force did Washington consider cooperative limitations. However, given the potential verification difficulties, any credible MIRV agreement proved exceedingly difficult. The MIRV story exemplifies the potentially self-defeating dynamics that decisive new applications of military technology may create, as what was initially perceived as a "killer application" for the United States rather quickly became Washington's greatest single concern in the late 1970s and 1980s. With the benefit of hindsight, it seems that if the same approach to ABM had been employed with MIRV, the United States and Soviet Union could have avoided another intense round of the arms race, though as MET would predict, it is unlikely that either state could ignore such a potentially decisive weapon in the context of their longstanding rivalry.

The remainder of this chapter will consider the history of the SALT I accords and the respective negotiations on ABM, delivery vehicles (ICBMs and SLBMs), and MIRV. Strategic talks were initially proposed by President Lyndon B. Johnson, the SALT I phase culminating with the May 1972 Moscow Agreements signed by President Nixon and Secretary Brezhnev. The first section will briefly discuss the security environment that served as the context for the SALT process. The second and third sections discuss the major weapon systems and the expectations developed by U.S. and Soviet military and political leaders, respectively. It will also consider the question of verification and compliance and the perceptions about the prevailing means of surveillance. The fourth section will briefly review the history of SALT under the Johnson and Nixon administrations in the context of MET. The conclusion will consider the hypotheses of MET against other relevant alternative hypotheses derived from theoretical and historical perspectives.

THE SUPERPOWER SECURITY ENVIRONMENT

The Johnson administration initiated discussions on strategic arms limitation in 1966, and pursued them with little success until just prior to Johnson's departure from office in 1969. The focus of the Johnson administration initiatives was a cooperative limitation on ABM defense systems, which a wide consensus of the military and political leadership had viewed as potentially decisive, but ultimately infeasible in a relevant time period.[2] U.S. leaders perceived the emerging competition in strategic defenses would lead to a new spiral in the arms race, the net result of which would only decrease the security of the two states at the expense of significant national resources.[3] Throughout the late 1960s, the United States approached the Soviets from a position of relative strength. Assuming office in the wake of a perceived "missile gap," the Kennedy and Johnson administrations focused resources on the development of a large, diverse strategic force, including a large intercontinental ballistic missile force that would guarantee a retaliatory capability under most conceivable Soviet building programs.[4]

The Nixon administration found a somewhat different context. By 1972 the Soviet Union had achieved parity with the United States in terms of ICBMs and threatened to open a significant quantitative advantage in offensive missiles.[5] While the Johnson administration had signaled interest in focusing primarily on ABM limitations, the Soviets had consistently responded with offers for a more comprehensive agreement that included both defensive and offensive weapons. By the early 1970s, it was the United States that was focused on the need for a comprehensive agreement, while the Soviets had moved to a more limited focus on ABM restrictions. In a relatively short period, the momentum of the arms race had seemingly shifted in the Soviets' favor. This situation presented the Nixon administration with the far more challenging task of negotiating both a limitation on ABM defenses and some cooperative agreement to maintain relative parity with the Soviets in terms of offensive strategic weapons.

Additionally, in geopolitical terms, by 1970 the highly bipolar world of the early Cold War had also seemingly shifted toward multipolarity, with Western Europe and Maoist China charting increasingly independent foreign policies.[6] These developments represented challenges and opportunities for both superpowers. The Peoples Republic of China's assertion of leadership within the world revolutionary movement and their geographic proximity to the Soviets presented a significant threat to Soviet security and diplomatic interests in a relatively short period of time.[7] Given the ongoing war in Vietnam, the Nixon administration attempted to capitalize on the Sino-Soviet split in order to alleviate the pressures on U.S. forces fighting on behalf of South Vietnam and also to put pressure on the Soviets to restrain its support for the North.[8]

The shifting international political environment, coupled with the central challenge of managing the arms race in the context of domestic political constraints against, thus shaped the diplomatic approach of President Richard Nixon and his National Security Advisor, Henry A. Kissinger.[9] While parity seemed to offer opportunities to improve diplomatic relations, the Soviet expansion exhibited no signs of slowing down, implying that Moscow may be attempting to achieve strategic superiority, thus undermining the notion of assured destruction and contributing to heightened mistrust and fear in the United States. The United States, accepting the perceived limitations on its own ability to compete in an uncontrolled arms race, clearly preferred that the Soviets embrace the concept of assured destruction and realize the futility and inherent dangers of attempting to achieve superiority.[10]

Thus, in contrast to the Johnson administration priorities, President Nixon and National Security Advisor Kissinger viewed offensive limitations as a mandatory component of any agreement.[11] Without the strategic offensive advantage that the Johnson administration had enjoyed, Nixon and Kissinger were forced to focus on offensive limitations in combination with a limitation on ABM. The roles had essentially been reversed. The Soviets, fearing the offensive advantage of the United States during the mid-1960s, were reluctant to engage in any cooperative arms control agreement that would leave the United States in a position of superiority.[12] By 1970, the Nixon administration was faced with a growing Soviet offensive force with no American expansion planned aside from the replacement of existing systems with the Minuteman III ICBM and Poseidon SLBM programs.[13] In this context, the SALT process was an attempt to manage the strategic arms race and avoid the potential dangers of unilateral policies that could decrease crisis and arms race stability.

U.S. MILITARY EXPECTATIONS

This section will evaluate the expectations relative to the military impact of the three major strategic programs considered in the SALT negotiations: ABM, delivery vehicles (ICBMs and SLBMs), and MIRVs. As the discussion of expectations relative to the military impact of respective strategic weapon systems will show, U.S. and Russian military and political leaders developed similar expectations regarding the impact to the three major strategic systems in question during the period. As MET would predict, their expectations shaped their willingness to consider limitations or reductions, and while the United States and Soviet Union could forge agreements limiting both ABM and delivery vehicles, limitations or prohibitions on MIRV proved impossible.

Antiballistic Missile Systems: Decisive but Infeasible

Antiballistic missile defense systems had been viewed as a potentially decisive application almost from the time that offensive missile emerged as a central application of the post-World War II strategic environment. The United States began experimenting with ABM technology in the mid-1950s.[14] However, from the time of Sputnik, U.S. energies focused around the development of a massive offensive ICBM force to complement the existing American superiority in manned, long-range bombers. Perceiving a "missile gap" that could threaten the bomber force on the ground, the Kennedy administration campaigned on and followed through with the expansion of U.S. strategic ICBM and SLBM forces during the 1960s. Having achieved superiority over Soviet offensive missile capabilities in a relatively short period of time, the mission to develop an effective ABM defense system lagged behind.[15] The Soviet Union seemingly invested in strategic defensive systems earlier on than the United States. The Tallinn line system, which was viewed as either a primitive ABM system or a more advanced air defense network, caused deep concern within the U.S. defense establishment, and ultimately served as a major driver of what would become the American MIRV program.[16]

The United States was able to delay deployment of ABM precisely because of the relative surplus of security provided by the offensive superiority of the Kennedy strategic buildup. The difficulties inherent in the development of an effective area defense capability were understood early on in the U.S. development program and influenced U.S. planners to focus on penetration aids in future offensive systems to overcome or overwhelm any strategic defenses that could be deployed. From a relatively early stage in the development of the U.S. strategic program, penetration of Soviet defenses was viewed as a generally more attainable goal than the defense of large areas of the continental United States from Soviet missile attack.[17]

By 1967, pressure had mounted within the Johnson defense department, particularly among the uniform military leadership to deploy a system, even if it was limited in scope.[18] Advocates of missile defense had been instigating for greater funding and a larger commitment to the various programs, but ultimately the decision to deploy came down to two distinct factors. Most importantly, the perceived Soviet decision to deploy ABM and the failure to achieve even limited agreements at Glassboro, created an intense pressure to keep up with a potentially decisive application that the Soviets had been working on for some time.[19] Secretary McNamara's January 1967 memorandum to the president laid out the major points:

> Our own anti-ballistic missile system, the Nike-X, has now reached a stage of development where it may be feasible to start concurrent production and deployment. . . . The Joint Chiefs have reaffirmed their recommendation that a decision be made now to deploy, with an initial operational capability in FY 1972, a Nike-X system which would provide for an area defense of the continental U.S. and a local defense of 25 cities against a "low" Soviet threat.[20]

At the time, McNamara and his principal civilian advisors in the Pentagon did not see the need to engage in a program directed at Soviet strategic forces because of the underlying commitment to an assured destruction capability and the negative potential implications of a U.S. deployment. He was also unmoved by the potential Chinese threat or the need to protect the Minuteman force, though the latter rationale would become a key concern in the Nixon defense department.[21] Although the protection of the U.S. land-based strategic offensive forces against the kind of heavy sophisticated missile attack the Soviets may be able to mount in the late 1970s might later prove to be worthwhile, it was still seen as premature to produce and deploy the Nike-X for that purpose.[22]

More importantly, for McNamara, the civilian leadership of the Department of Defense and the majority of the scientific advisors to the Johnson administration, the deployment of ABM would not improve the U.S. position and could only have the impact of undermining the system of assured destruction and stable deterrence between the superpowers. For their part, the Joint Chiefs (JCS) supported the president's attempts to engage the Soviets in cooperative limitations on ABM, but as the initiatives proved unsuccessful, the JCS advocated for deployment to avoid falling behind the Soviets as well as addressing the emerging smaller-scale threats.[23]

When Secretary McNamara announced the decision to deploy the Safeguard ABM system, on August 1967, it was presented as being a "thin" system, focused on the emerging Chinese threat, the problem of accidental launches and possible blackmail attempts. While it was a reaction to the unwillingness of the Soviets to constructively address the issues of arms control at Glassboro, the deployment was couched in language that avoided directly engaging the Soviet Union precisely because Johnson and his advisors believed the potential may still exist for an arms control agreement before the end of his term.[24] The invasion of Czechoslovakia effectively precluded this possibility.[25]

As discussed above, the Nixon administration faced a different security environment when it assumed office in 1969. The Soviet buildup had effectively succeeded in achieving a level of numerical equality with the strategic offensive forces of the United States and now threatened to surpass U.S. force levels.[26] More troubling for the Nixon defense department, the United States had essentially frozen its force levels in the mid-1960s. No new programs were expected to be deployed until the MIRV-equipped Minuteman III ICBM and Poseidon SLBM programs in the mid-1970s. These represented major qualitative advancements because of the MIRV technology, but now even the successful deployment of these programs would not close the perceived widening gap between U.S. and Soviet forces in quantitative terms. U.S. planners feared that if the Soviet building programs continued, the sheer numerical advantage could actually undermine the credibility of the U.S. deterrent force and reintroduce the possibility of a first-strike or damage limitation logic into the superpower strategic relationship. Thus, after assessing the state of U.S. force posture and readiness at the outset of the administration, the Nixon administration opted to transform the thin, Chinese-focused Sentinel ABM program into Safeguard.[27] While remaining a relatively limited ABM deployment, military planners viewed the fundamental mission of Safeguard as maintaining the credibility of the U.S. deterrent posture by protecting the land-based ICBM component of the strategic triad against possible Soviet attack.[28] The Soviet threat to the U.S. land-based force necessitated the deployment of Safeguard to address the growing quantity of Soviet missiles and the even more disconcerting possibility of Soviet adoption of multiple-warhead technology.[29]

Thus the mission of Safeguard was still more limited than a widespread "area defense" capability meant to protect the continental United States from an all-out Soviet attack, as more ambitious proponents of missile defense systems advocated. At the same time, it is important to note that Safeguard faced significant domestic popular and Congressional opposition and adequate funding for the program would prove difficult.[30] However, in the view of the Nixon administration, the deployment of Safeguard had now become essential to the maintenance of a credible U.S. deterrent and was thus perceived as necessary to the maintenance of assured-destruction capabilities by the United States.[31] The perception of a consistent Soviet commitment to develop what seemed to be a more ambitious ABM defense of Moscow also contributed to an underlying concern among decision makers that the Soviets could achieve a breakthrough in ABM that would dramatically change the existing system. By 1970, the Soviets had been working on the Moscow system for almost five years and intelligence estimates based on satellite surveillance reinforced the perceptions among the Nixon administration that the Soviet efforts had progressed. In a very real sense, this momentum, when coupled with the relatively minimal U.S. efforts in ballistic missile defense during the later 1960s, created a perception that the United States had fallen behind, and that the implications of such a gap could be decisive.[32]

The discussion of ABM under the Johnson and Nixon administrations therefore reflects a broad consensus among leading decision makers.[33] ABM was viewed as a *potentially* decisive, even revolutionary, military application. However, it is clear that the technological barriers to achieving an effective system, even against more limited threats such as the Chinese strategic program or accidental missile launch, were daunting and well understood. While the United States maintained a relative advantage over the Soviet Union in offensive strategic systems, the prospect of a Soviet ABM was viewed as less threatening. Policymakers remained confident that any such system could be overwhelmed by American strategic assets. However, as the Soviets achieved numerical equality with the United States and threatened to open a quantitative gap in offensive systems in their favor, this confidence evaporated and fears of a Soviet advantage couple with a credible strategic defense undermined the perceived credibility of the U.S. deterrent posture. The Soviet buildup forced the United States to respond. Given the effective freeze in U.S. building of offensive forces from the mid-1960s on, the perception of the importance of developing and deploying an ABM system that could at the very least provide some insurance against a first-strike on U.S. land-based assets was increasingly attractive to U.S. policy makers. This does not imply that decision makers in the Nixon administration held more optimistic views as to the feasibility of an ABM system. Safeguard remained a relatively limited program in scope, like its predecessor Sentinel. However, its primary mission, the protection of the Minuteman force, reflected a change in the strategic environment created by the Soviet buildup in both strategic offensive and defensive forces. Thus, we would expect that the cooperative limitation of ABM systems would remain a diplomatic priority of the United States under the Nixon administration, as it had been under President Johnson. However we would also expect to see a greater emphasis on quantitative offensive limitations to address the perception of growth in Soviet forces toward levels that could make first-strike conceivable.

Ballistic Missiles: Land-based ICBMs and SLBMs

Since the Soviet launch of Sputnik in 1957, the nature of the arms race increasingly became dominated by the respective missile forces of the superpowers. The fears of a Soviet advantage in intercontinental ballistic missile capabilities necessitated a buildup in this class of weapon by the United States as well as other assets to provide a diverse, comprehensive deterrent capability.[34] The foundation of a strategic triad of weapons, including manned bombers, ICBMs, and submarine launched ballistic missiles would define the American deterrent posture. In fact, by November 1961, the fears of the missile gap had been dispelled, after the expansion of strategic offensive systems under the Eisenhower and Kennedy administrations was well underway.[35]

After the initial buildup under the Kennedy administration, it became increasingly accepted by the consensus of U.S. military and political leaders that once a level of assured destruction had been reached, subsequent increases would prove of little military value.[36] To a large degree, this is reflected in the fact that the United States did not engage in any new building programs after 1967.[37] Concern remained for qualitative improvements in existing forces, and maintaining those quantitative levels deemed necessary to maintain a credible deterrent, but after the initial buildup, there was little perceived need to continue expansion.[38] As discussed above, this contributed to the deterrent posture of the United States and thus provided a level of confidence in U.S. capabilities despite a perceived Soviet buildup in the late 1960s.[39]

Underlying this analysis was the view that missile technology had, on most important dimensions, effectively achieved relative perfection.[40] The solid-fueled missile programs that would comprise the U.S. strategic force after the mid-1960s, the Minuteman and Polaris were viewed as more dependable, survivable, and cost-effective than the older Atlas and Titan missiles which were phased out of the arsenal.[41] While the late 1950s and early 1960s saw significant leaps in the development and refinement of ICBM and SLBM technology in terms of payload, accuracy, and reliability, by the 1970s, this process had slowed.[42]

> Minuteman has steadily evolved as new components and new operational concepts have been developed since 1960. The Minuteman III now (1070) being deployed will hurl a larger payload a longer distance and will do so with much finer accuracy than the original version of this weapon.[43]

Qualitative improvements could be made in terms of accuracy (the circular error probability or CEP), throw-weight, or the volume of the payload of the missile and reliability. However, these were generally viewed as incremental improvements, and ultimately minimal when considered in relation to the quantitative levels of strategic forces. This process of development was similar for SLBM. "The A-3 [Third generation of the Polaris SLBM], as in the case of the later Minuteman, had a longer range and a larger payload than the first," but ultimately, these assets were only incremental improvements over their predecessors.[44] The missile revolution of the late 1950s had been replaced by the delivery vehicle evolution of the 1970s. At the extreme margins, larger throw-weight and higher megatonnage or higher levels of accuracy could contribute to the counterforce potential of a given missile, but such arguments also implied the quantity of missiles engaged under such a scenario.[45] In fact, by 1969, the Chief Negotiator of the U.S. SALT delegation, Gerard Smith would remark that "with the exception of one or two uniformed military officers all believed our ICBMs were something of a wasting asset."[46] Similarly, Herbert York, who had contributed to development of the U.S. strategic arsenal in variety of capacities in his career argued that Poseidon "looked like nothing more than a fourth version of Polaris that had been given a new name."[47]

The prevailing view was that missile technology had, on most important dimensions, effectively achieved perfection. The solid-fueled missile programs that would comprise the U.S. strategic force after the mid-1960s, the Minuteman and Polaris were viewed as more dependable, survivable, and cost-effective than the older Atlas and Titan missiles which were phased out of the arsenal.[48] Thus, in a rough parallel to capital ships of the 1920s, delivery vehicles—both hardened ICBMs and SLBMs—were viewed as relatively known commodities. As such, the possession of these assets in some quantity deemed necessary for a credible deterrent posture made continued numerical expansion of little value in military or security terms.[49] These weapons served as the basis for the concept of assured destruction and the U.S. deterrent posture and provided U.S. leaders with a level of confidence in the face of the Soviet buildup in the mid-1960s. To maintain the security of both states and to perpetuate the seeming stability of mutually assured destruction, agreed numerical limitations on these weapons could contribute to an arms control agreement. While deeper cuts or reductions in forces may have been an ideal objective of more ambitious arms control, given the existing relationship between the superpowers mutually agreement on force levels would provide a first-step that would maintain security without enhancing risk. This was the position of the early attempts at SALT under the Johnson administration.[50]

National Technical Means of Surveillance: Supporting Negotiations

The challenges of U-2 high altitude overflights of Soviet territory had proven dangerous. New satellite technology would prove invaluable to the development of robust means of surveillance of the Soviet Union and its military deployments. Specifically, the Corona satellite program would constitute a breakthrough for U.S. national security and intelligence gathering.[51]

> This new capability represented a true revolution. Intelligence collection by reconnaissance satellites improved enormously American knowledge of Soviet military capabilities and economic strength: "Every square inch of Soviet territory was opened to American eyes."[52]

By the late-1960s, the U.S. intelligence community could report to the president and his military and defense advisors with relatively high degrees of certainty about weapons programs in the Soviet Union. In preparation for the initial SALT discussions, the interagency SALT steering committee concluded that most Soviet programs could be monitored with relative confidence, certainly before any breakout or expansive quantitative advantage could be achieved.

> An intelligence estimate concluded that there are aspects of potential agreements that can be verified by national means of intelligence with high confidence. While some Soviet violations could get underway without our knowledge, detection in most cases would be assured prior to the creation of a substantial Soviet buildup.[53]

In considering a "Stop Where We Are" proposal to the Soviet, the intelligence community similarly concluded that U.S. national technical means of surveillance, primarily based upon the Corona satellite program, would indeed provide the requisite coverage of Soviet strategic programs to provide confidence that cheating would be detected.

> The CIA estimates that a SWWA agreement could be verified with high confidence. Because all strategic systems would be constrained under SWWA, there would be theoretically be a broad spectrum of possible types of violations.[54]

Given the nature of Soviet programs, offensive weapons systems like ICBMs and missiles of lesser range, SLBMs and submarines, and bombers could be monitored with relative ease. Similarly the components of a relatively large-scale ABM system, including various radars and detection installations, and interceptors could be viewed through satellite imagery without significant fears. Unsurprisingly, because of their size, MIRV warheads would present a much more difficult verification challenge.

> In the event of a moratorium, we would have confidence in our ability to monitor within narrow limits the number of ICBM sites and ballistic missile

> submarines. Monitoring MIRV/MRV testing is a more difficult problem and is one of those matters now under intensive review.[55]

So while large-scale flight-testing of warheads could potentially be tracked and discerned from national technical means, the actual production, development and deployment of MIRVs that were somehow illicitly tested would leave the United States unaware and potentially vulnerable in the event that the Soviets deployed MIRV in large numbers on existing or newly deployed delivery vehicles.

> Notwithstanding the net judgment of Intelligence Community, some argue that our assurance in verifying a MIRV prohibition through a ban on MIRV flight-testing could be uncertain.[56]

On-site inspection would likely be necessary to effectively verify any MIRV-ban agreement.[57] Given the nature of the Soviet regime, the idea of Moscow granting intrusive on-site inspections on Soviet soil to U.S. inspectors seemed impossible and would likely prohibit any such agreement. However, given the decisive nature of MIRV technology, it is unsurprising that such a high threshold was set for verification. The perceived value and benefits of MIRV are discussed in the next section.

MIRV Technology: Decisive (and Highly Threatening)

Multiple independently targetable reentry vehicles were viewed as a decisive breakthrough in military technology by U.S. military and political leaders. It represented a technological breakthrough in the missile age that offered to provide the United States with a capability that could maintain U.S. security and the credibility of the U.S. deterrent in the context of a variety of Soviet deployment scenarios. It was truly perceived as a "killer application" by decision makers in the Kennedy and Johnson administration. MIRV emerged as a solution to the potential problem of a large-scale Soviet ABM system in the 1960s. After experimentation with a variety of possible penetration aids to ensure that U.S. missile would be capable of delivering a massive retaliatory strike in the event of a Soviet first-strike, the concept of multiple warheads emerged as the most robust and effective response.[58]

Planners in the McNamara defense department were skeptical of the value of quantitative additions to the American strategic missile force beyond the force levels achieved during the initial Kennedy administration buildup.[59] The resources necessary to continue to expand strategic offensive forces would preclude investment in other areas that the administration had viewed as important, and given the lack of any perceived military or security benefit provided by such an expansion in the numbers of missiles, MIRV offered an important capability. Even if the Soviets succeeded in a breakthrough on

ABM defense, the United States would continue to possess an effective deterrent based upon the ability to penetrate those defenses offered by MIRVed delivery vehicles.[60] This capability further allowed the United States to forego the rushed deployment of its own ABM system, and commit funding only to research and development activities, hedging against a possible breakthrough and targeting resources more effectively. Military and civilian leaders in the Pentagon were highly sensitive to the protection of the MIRV program and were adamant about excluding its discussion for any potential arms limitation discussions.[61]

Additionally, the nature of MIRV development was such that, as opposed to ABM and delivery vehicles, defense and intelligence experts were not optimistic about detecting Soviet MIRV programs.[62] Thus, precisely because of its decisive nature and because of the difficulties associated with surveillance, most civilian and military officials were opposed to any limitations on MIRV. As JCS Chairman General Wheeler wrote in a memorandum to Secretary McNamara in January 1967:

> In contrast to this, it is proposed in the paper that, since unilateral intelligence is incapable of detecting changes incorporated in missile to improve their performance, no limit be placed on changes in missile characteristics as such. The Joint Chiefs of Staff believe that, in order to avoid Soviet technological surprise and resulting degradation of US military posture, research, development, and prototype testing must be permitted under any freeze agreement and that US-planned improvements in its missile characteristics must be pursued.[63]

While generally supporting the efforts of the Johnson administration to negotiate a successfully arms control agreement with the Soviet Union, particularly with respect to ABM systems, the JCS as well as the civilian leadership of the defense department viewed MIRVs as a decisive weapons system and thus vehemently opposed their limitation in any form under a prospective agreement. So decisive was MIRV technology that even the initiation of discussion on MIRV was effectively taken off the table.[64] Because of the centrality of MIRV technology to the future of the U.S. strategic arsenal, diplomatic initiatives should refrain from including any restraints or limitations on MIRV development. To prevent the Soviets from moving in the direction of MIRVs, measures were taken within the Department of Defense and across the administration to maintain its secrecy.[65]

Throughout the mid to late 1960s, the policy of the Johnson administration with regard to arms control with the Soviets accepted this position, underscoring the decisiveness with which the consensus of defense planners and decision makers viewed MIRV technology to the U.S. strategic capability. Instructions for potential negotiations with the Soviets drafted in July 1967 made this point clear.

> No additional restrictions would be imposed upon technological improvements of launchers or missiles already deployed including increasing the hardness of silos and deployment of MIRV's, or the retrofitting of existing launchers with new missiles.[66]

Over time, discussion about including MIRV in comprehensive negotiations with the Soviets emerged and created some controversy. Precisely because of the decisive nature of the technology, advocates of more intrusive arms control and critics of the Johnson administration argued that testing and development of MIRV should be restrained in advance of strategic arms limitation talks with the Soviets.[67] By the time the SALT negotiations took place, the commitment to develop and deploy MIRV was predetermined and unaffected by the ebb and flow of arms control negotiations. MIRV would be a decisive application for the U.S. strategic force. Conversely, it was also an application that would be difficult to cooperatively restrain under an arms control agreement. Moreover, while the other aspects of SALT initiative were predicated on the ability of the United States to effectively monitor Soviet strategic programs on the basis of existing national means of verification, MIRVs would prove far more difficult. A Special National Intelligence Estimate delivered in November 1968 highlighted the inherent problems associated with assessing Soviet progress in the field of multiple warheads.

> We believe, for example, that we could detect and identify Soviet testing of multiple independently-targeted reentry vehicles (MIRVs) for intercontinental ballistic missiles. We could probably also detect activity associated with an antiballistic missile (ABM) system. . . . We see no prospect of determining whether MIRVs (if developed) or other significant improvements had been incorporated in deployed offensive missiles.[68]

Considered together, the decisiveness nature of MIRV technology and the perceived difficulties in assessing the progress of Soviet deployment of MIRV over time effectively precluded the ability of the United States to engage in any discussion of their limitation.[69]

Ultimately, the MIRV program did not prove to be a perfect hedge against Soviet behavior. The rapid quantitative buildup of Soviet offensive forces and the expansion of the Soviet ABM system increased U.S. fears concerning the credibility of the U.S. deterrent over time. However, this only made MIRV more central to the maintenance of the U.S. assured destruction capability.[70] In hindsight, it seems somewhat puzzling that the United States would have been so confident in the security provided in the deployment of MIRVs given the action-reaction cycle and the expectation that the Soviets would likely not be far behind in deploying similar capabilities. In such an event, the confidence of the previous estimates of a world in which only the U.S. possessed MIRVs would seem to be seriously undermined. Most importantly, the massive size of the Soviet SS-9 now presented an acute threat should the Soviets achieve MIRV technology because of the potential multiplication of deliverable warheads that a large MIRVed SS-9 force could represent. In the context of the growing quantitative Soviet offensive threat and the Soviet ABM system, the Nixon administration had reason for concern.[71]

The prospect of Soviet MIRVs would only exacerbate what was already a fairly threatening strategic environment for U.S. defense planners. At the same time, MIRV represents a textbook case of MET in that the application was deemed so decisive that military and civilian leaders could not fathom foregoing the capabilities that it would provide. The deployment of MIRV would unilaterally improve U.S. capabilities in a dramatic way in an arms race environment. Subsequent Soviet acquisition of MIRV would remove this advantage or, perhaps, shift the balance in favor of the Soviets in combination with other activities on both sides. What is important to understand is how the expectations relative to the impact of MIRV drove the behavior of military and civilian leaders, even perhaps when limitation of MIRV should have (in hindsight) been seen perhaps a more rational, forward-looking option.

SOVIET PERCEPTIONS AND EXPECTATIONS

The Soviet Union shared many of the same expectations about the key weapon systems under consideration for restraint during the SALT talks, though they had not necessarily taken the same path to reach these conclusions.[72] For much of the early Cold War period, Soviet military planners were singularly concerned with the task of fighting and winning a nuclear war.[73] The perceived vulnerability of the Soviet Union exposed during the Cuban Missile Crisis forced the Kremlin leadership to engage in a Soviet buildup, particularly in strategic offensive missile forces after Khrushchev's ouster.[74] However, by the mid-1960s, it seemed that the Soviets were undergoing a process of reevaluation of nuclear doctrine and strategy similar to that of the McNamara defense department after the initial U.S. strategic buildup.[75] Soviet political and military leaders had generally come to the conclusion that nuclear war was not inevitable and that the stability provided by nuclear deterrence and the underlying relationship of assured destruction could provide a basis for peaceful coexistence.[76] Such a doctrinal shift would seem to contribute to the likelihood of cooperative arms control, particularly in the context of relative strategic parity between the two superpowers.[77] However, the Soviet expectations relative to the strategic systems under discussion played the central role in shaping Moscow's willingness to engage in SALT.

Soviet ABM

By 1970, Soviet defense planners had come to a similar view of those in the United States that ABM systems, while possessing the potential to provide dramatic defensive capabilities, were unlikely to do so in any relevant time frame.[78] The Soviet emphasis on ballistic missile defense, exhibited by the deployment of the Moscow ABM system contributed to the war-fighting mission by ostensibly providing a defense capability around the national command or perhaps larger strategic areas within the Soviet Union.[79] In fact, the Soviets had spent more than a decade working on the problem of ballistic missile defense and had little to show for it.[80] Despite its fairly large scale, the deployment of the Galosh system around Moscow had failed to achieve the necessary thresholds of effectiveness to credibly provide for the protection of the Soviet capital.[81]

Moreover, by this time, the Soviets were actually concerned that the trends of technological competition had shifted in the favor of the United States. The Sprint missile system that had been developed to be utilized in the Sentinel (and then Safeguard) systems was seen to have greater potential capability and technical sophistication than the Galosh rocket, which had proven fairly ineffectual.[82] Similarly, the Soviets were unenthusiastic about the prospects of the effectiveness of the radar systems that had been deployed.[83] In sum, Russian military leaders did not view the probability of deployment of an effective large-scale, area defense system to be technologically feasible in the short to medium term. Moreover, their more extensive experimentation and development activities provided the Soviets with a clear view of the probability of success in this endeavor and thus they were increasingly supportive of a cooperative means of ending or limiting the competition with the United States over strategic defense.

Soviet Strategic Offensive Forces

Quantitative expansion of the Soviet ICBM force had been the centerpiece of the strategic buildup in the post-Khrushchev era.[84] The Soviets viewed the objective of closing the gap between Soviet and U.S. strategic forces as essential to maintaining their capacity to fight and win a nuclear war with the capitalist powers.[85] While deterrence became more widely considered to be the appropriate primary objective of Soviet strategic forces, those forces continued to reflect the war-fighting emphasis during the 1960s.[86] For example, the 280 massive SS-9s, whose size had initially puzzled U.S. analysts, were built to perform the mission of disabling the United States land-based ICBM control centers.[87] With its 20-25 Megaton (MT) payload, Soviet planners viewed the likely impact of these large-throw-weight ballistic missiles as providing the potential to cripple the U.S. land-based ICBM force in a likely preemptive first-strike scenario.[88] The U.S. bomber force, Minuteman silos, and soft targets would be addressed by the 720 smaller 1-MT SS-11, which was the staple of the Soviet arsenal during the mid-1960s buildup. Its smaller size and cost allowed for its procurement in greater numbers and it was a more versatile system, devoted to both ICBM and IRBM missions against targets in Europe.[89]

Having traditionally focused on its geopolitical status as a major landpower, the development of a larger submarine force and SLBMs was a relatively new capability in the Soviet arsenal.[90] However, the perceived advantages of these assets in terms of their ability to contribute to the fighting of a nuclear war and, over time, to the maintenance of a credible second-strike capability, made the construction of a large submarine force a priority in the late 1960s.[91] SLBMs and submarine provided the benefit of being survivable and relatively cost-effective assets and having lagged behind the United States, a traditional maritime power with significant overseas bases, it was not surprising that Kremlin planners sought to deploy a large number of nuclear submarines.[92] This was an area where the Soviets could, through concerted effort and investment, significantly rectify an existing deficit in strategic weapons and potentially overtake the perceived U.S. advantage.[93]

Like their U.S. counterparts, the Soviets viewed missile technology as "taken for granted" by the time of the SALT conference. While ICBM represented a technological breakthrough in the late 1950s, providing the capability to visit massive firepower upon a target half a world away and with no real defense, by the 1970s this was an accepted fact of life for both superpowers.[94] Possessing these assets in some number to provide a guaranteed retaliatory strike was the only real concern for planners in the Kremlin, as it was for their counterparts in Washington.[95] Qualitative improvements would be made over time to improve accuracy, payload, and reliability, but the underlying technology was understood and accepted, and Soviet technology had proven to be able to compete with that of the United States.

MIRV: Short-Term Concern, Long-Term Opportunities

The question of MIRV is somewhat more difficult to address in the concept given available Soviet literature. However, it seems clear "Soviet planners were acutely aware of the significance of MIRV technology."[96] The United States was indeed quite far ahead of the Soviets in terms of independently targeted warheads. The Soviets had experimented with a "triplet" of warheads for the large SS-9, and this, to a large degree, raised concerns about the security of the Minuteman force and pushed toward the deployment of Safeguard.[97] However, the triplet was an early experiment in MRV (Multiple Reentry Vehicle) technology, without the crucial (and decisive) ability of independent targeting. The successful deployment of the triplet would represent an improvement, but far less important than a true MIRV capability which offered the potential to deliver warheads to a number of individual targets and thus truly multiply the potential capability of a single delivery vehicle. Nonetheless, MRV capability was typically considered a precursor to the development of an effective MIRV capability, and thus while the Soviet lagged behind the United States in the development of this decisive military application, they were on the path. At the same time it was a further hedge that along with the Soviet quantitative advantage could maintain the Soviet deterrent until MIRV deployment.[98] Ultimately then, it seems highly unlikely that the Soviets would have agreed to lock themselves into a position of inferiority in this important technology, particularly when it was exhibited as technologically feasible.[99] MIRV, quite simply, was too tempting to forego and too frightening to ignore.

The Soviets indicated an interest in discussing a MIRV limitation, particularly given their position behind the United States in this important development, as MET would predict, they were unable to offer or agree to an effective limitation scheme, precisely because they were unwilling and unable to lock themselves into a position of inferiority vis-à-vis the United States. As Dobrynin recalled, "In a private talk Kissinger bantered with me, 'Well, you are very smart indeed to suggest banning something you don't have and we already do.'"[100] Moreover, the eventual U.S. offers to restrain MIRV at Geneva would be rejected by the Soviets precisely because of their intrusive requirements in terms of verification and compliance, and the Soviets were not eager to restrain their ability to develop MIRV technology for their own strategic forces.

Military and political leaders in the Soviet Union had thus, by 1970, come to similar conclusions as their counterparts in the United States in terms of the perceived value of strategic offensive and defensive forces and the doctrine under which they would be employed. After the achievement of quantitative force levels to guarantee a massive retaliatory strike that would be unacceptable to the United States, the Soviets gradually understood the war-fighting mission of strategic forces as secondary at best to the more pressing concern of deterrence. The failure to achieve an effective ABM system only contributed to this understanding and realization that fighting a nuclear war was essentially irrational given the inability of either side to develop means to survive such a calamity. This does not mean that war-fighting was completely removed from the calculations of either superpower, but only that both viewed the relative strategic stability of a relationship of assured destruction as a positive effect of the arms race. Taking efforts to cooperatively restrain or limit unilateral activities that could undermine that stability thus took on a new priority. At the same time, this acknowledgment did not remove the pressure to continue to maintain robust research and development capabilities and thus limit the potential for cooperative restraints on quantitative advances to existing forces. This is reflected in the expectations of MIRV technology as a decisive advancement in the arms races and would contribute to the inability of either superpower to develop cooperative means to restrain it over time.

THE STRATEGIC ARMS LIMITATION PROCESS

The groundwork the SALT I accords, which included the ABM Treaty and the Interim Agreement of Offensive Weapons, signed by President Richard M. Nixon was laid down during the last years of the Johnson administration. However, by the time President Nixon signed the agreement with Premier Brezhnev in May 1972, the underlying political and military contexts were significantly different.

Toward SALT: The Johnson Administration

The administration of President Lyndon B. Johnson initiated the exploration of strategic arms limitation talks with the Soviets in the mid-1960s. President Johnson, Secretary of Defense Robert S. McNamara, and Secretary of State Dean Rusk strongly supported the search for a cooperative solution to the central problem of the strategic arms race: antiballistic missile systems. Since the Cuban Missile Crisis, and during the ensuing years, a consensus had developed across the political and military leadership of the United States that national security could be effectively guaranteed through the employment of a doctrine of nuclear deterrence, predicated on the concept of assured destruction. Having achieved a significant advantage over the Soviets in terms of strategic offensive forces in response to fears of a perceived missile gap at the outset of the Kennedy administration, by the mid-1960s, the United States was fairly secure. It had achieved a level of strategic offensive armaments that guaranteed a second-strike capability against any Soviet attack, thus creating an environment in which the initiation of nuclear conflict was inconceivable. Thus the United States had attained an assured-destruction capability, deemed essential to deterring a Soviet first-strike. It also provided the United States with the perceived leverage to forge a bargain with the Soviets.

Moreover, the experience of the "missile gap" controversy, which had emerged during the 1960 presidential campaign, had influenced the key advisors of the Kennedy and Johnson administrations to view the strategic arms race as an action-reaction cycle, in which actions by one superpower inevitably influenced the behavior of the other.[101] As the missile gap episode seemed to show, however, the reaction could often be greater than the initial action required, and over-shooting was a likely result of the uncertainty inherent in the superpower relationship. The deployment of an ABM system would likely spur a similar move by the Soviet Union. At the very least, the Soviets could be expected to deploy greater numbers of offensive weapons to overcome the perceived threat to their deterrent capability. In combination, these two actions could seriously undermine the U.S. assured destruction capability and undermine its deterrent posture. Thus the cooperative limitation of ABM could remove a serious threat to the emerging stability of the superpower strategic relationships and allow the states to avoid another intense and costly spiral in the arms race. Unfortunately, the Johnson administration would fail to achieve cooperation with the Soviets, but its efforts would set the stage for its successor. Moreover, as the Soviets achieved parity in the early 1970s, they were more receptive to comprehensive arms control, including both offensive and defensive systems, than they had been in the mid to late 1960s.

The central concern of defense officials in the Johnson administration, and the only threat to the American assured destruction capability, was the possibility of a Soviet breakthrough on ABM systems. Both the United States and the Soviets had been working on the development of ABM since the 1950s. Given the failure of the United States to develop an effective system, the administration considered a Soviet breakthrough unlikely, but as long as the Soviets continued to commit resources to the development of an effective ABM system, the possibility existed that a success could be achieved. Thus, while the technological solution to the challenge of defense against ballistic missiles remained elusive, the Johnson administration was pressed to continue to develop and perhaps deploy a small-scale ABM system to lay the groundwork for future improvement. In considering the deployment of a U.S. ABM system, the Johnson administration debated the likely Soviet response, and assessed the likely net impact on U.S. security of such an action to be negligible at best. The inability to develop a credible missile defense thus spurred the administration toward a diplomatic initiative to prohibit missile defense systems and thus avoid the possibility of a new phase in the arms race and maintain the existing strategic status quo. Despite the perceived practical difficulties involved in the deployment of an American ABM system, and knowing that the Soviets had experienced difficulties in the development and deployment of their missile defense capabilities, U.S. deployment was widely considered a major step in the arms race.[102]

While initiating a new phase of the arms race troubled many of the leading decision makers in the Johnson administration, the actual perceived military value of the system also caused concern. The achievement of an assured-destruction capability through the acquisition of offensive weapons seriously undermined the perceived value of an effective U.S. ABM system in the short term, particularly if the deployment of such a system would only lead to the logical reaction of a Soviet increase in offensive forces, the deployment of Soviet ABM, or some combination of the two.[103]

Building upon the concepts of assured destruction, deterrence, and the action-reaction model of the superpower arms race, by the mid-1960s the Johnson administration held the view that the arms race had effectively achieved a "plateau."[104] While the Soviets remained behind the United States in terms of numbers of offensive weapons, the superpowers were approaching a relationship of mutually assured destruction (MAD), where technology was unlikely to provide states with capabilities to overturn the deterrence-dominant system, and thus stability was possible. This would certainly be the case if ABM could be cooperatively restrained. The superpowers could thus avert a new cycle in the arms race, and the destabilizing political consequences of such an occurrence, through the cooperative prohibition of strategic defenses. Moreover, it was the prevailing view that another round of competitive deployments would be self-defeating and an agreement would allow both superpowers to "avoid pursuit of a higher plateau in the nuclear arms race which would not increase our security."[105] Nevertheless, if such an agreement proved unattainable, the erosion of relations between the superpowers was also viewed as highly probable. While the ultimate effect of another round in the action-reaction process of the strategic arms race may do little to improve the relative security of either power, the subsequent increase in tensions and suspicion could increases the probability of a conflict. Thus the likely political implications of unrestrained racing were also seen as advocating for a cooperative agreement with the Soviets. As CIA Director Richard Helms wrote to National Security Advisor Walt Rostow:

> a new surge of competitive arming would tend to sustain tension and mistrust between the two counties, and thereby to limit even more the possibility of moving our relations gradually into more constructive channels.[106]

In the larger context of the Cold War, the arms race had become a central component of the superpower rivalry and a major new deployment by the United States would have significant ramifications for Soviet policy and thus perpetuate a dynamic of competition that could undermine the perceived stability of assured destruction.[107] Thus, the consensus of the military and political leadership of the Johnson administration viewed the timing of negotiations toward a potential strategic arms limitation treaty as particularly appropriate. Such a process offered the potential to develop a cooperative solution to the destabilizing challenge of ABM and therefore contribute to the maintenance of the underlying deterrent-dominant security environment which could facilitate improved diplomatic relations between the superpowers.[108]

Unfortunately for the Johnson administration, the Soviet Union viewed the prospects of cooperative arms control, particularly the limitation of missile defense programs, with far less urgency than the United States. In the first place, the Soviets had deployed an ABM system, utilizing the Galosh interceptor rocket, around Moscow in November 1966.[109] This bolstered the so-called Tallinn Line, which had caused a great degree of confusion and consternation within the U.S. defense and intelligence communities because of its perceived application as an air defense or perhaps a missile defense system.[110] The Soviets had seemingly exhibited a firm commitment to the deployment of a large-scale ABM system, and this was troubling to U.S. planners. However, the Soviets were more troubled by the perceived U.S. advantage in strategic offensive forces, most notably its far superior long-range bomber force and ICBMs. Thus, it became clear that the Soviets were less than enthusiastic about American overtures that focused primarily on the question of restraints on ABM. The Soviets desired a more comprehensive discussion of armaments that would potentially include limitations on offensive weapons as well. U.S. superiority in strategic offensive weapons was the primary concern of the Soviet Union, and the Galosh system and quantitative improvements in Soviet ICBM forces were viewed as means to close the perceived gap with the American arsenal.[111]

The United States was clearly less concerned about the offensive side of the strategic equation precisely because of its advantage, but nonetheless signaled a willingness to engage in more negotiations that could lead to a more comprehensive agreement. After the Soviets rebuffed initial discussions that focused only on ABM restraints, Thompson adjusted the approach to include other classes of weapons, as the Soviet requested. The prospective deployment of a U.S. ABM system was viewed as the primary bargaining leverage to facilitate Soviet responsiveness.[112] However, an exclusive focus on ABM was unsatisfactory to the Soviets. The incorporation of offensive systems within a more comprehensive formula of restraints and limitations was viewed more likely to receive a favorable response. The Soviets were likely concerned that the focus on ABM would only limit their ability to defend against the (much larger) American missile force, while providing no restraint on UN strategic offensive programs. In reality, the United States had effectively stopped building, with their last planned Minuteman II missiles deployed in 1967.[113] The Soviets would want reassurance against U.S. quantitative expansion and the ability to fulfill a portion of the offensive strategic forces that were under development at the time.[114] Thus the rough outlines of a mutually acceptable proposal were under consideration by early 1967. Thompson, Rusk, and former Ambassador Charles Bohlen were optimistic about an agreement with the Soviets, though it would necessarily have to incorporate both offensive and defensive strategic weapons.[115] As Thompson wrote after delivering Johnson's initial message, "On balance, however, I believe we have a good chance of negotiating an ABM, ICBM freeze."[116]

However, the Soviets continued to resist engagement in formal negotiations. Even the direct intervention of President Johnson failed to bring out concrete action on the part of the Soviet leadership. The Soviet announcement of the Galosh system only added to the impetus to deploy an American ABM system, and Johnson attempted to use the growing domestic political and bureaucratic pressures toward deployment to attempt to push the Soviets to negotiate.[117] But while embracing the language of arms control and cooperation, the Soviets nonetheless failed to engage in any meaningful diplomatic discussion on the topic. Instead, they seemed far more concerned with taking unilateral measures to close the perceived gap between Soviet and U.S. offensive strategic forces.[118]

In reality, as MET would predict, Soviet behavior during this period seemed focused precisely on the goal of achieving a balance of power with the United States in terms of offensive nuclear weapons. Even the June 1967 summit at Glassboro, New Jersey, failed to move the dialogue forward.[119] Kosygin proved unwilling to make a firm agreement on ABM restraints and dismissed the idea of focusing on defensive systems. This rejection led directly to President Johnson and Secretary McNamara's decision to announce in September 1967 the deployment of a "thin" ABM system that was directed toward the threat of Chinese ICBMs and did not propose to provide a credible defense against a larger-scale Soviet offensive strike.[120] The decision to deploy the "Safeguard" ABM system represented a reversal for the Johnson administration and for the McNamara Defense Department. In reality, the decision reflected a compromise between Congressional advocates who sought a larger system, military and civilian experts in the Pentagon that viewed a smaller program and technologically feasible and thus warranting of deployment to further exploit a new and potentially decisive application, and those who opposed deployment. However, despite depictions of the decision as a bureaucratically driven process, in reality, the Soviet repudiation of President Johnson's overtures and the failure to achieve a substantial agreement at Glassboro, was the central precipitating factor. If the Soviets were indeed committed to the development of strategic missile defense, then the United States was more or less forced to engage in research and development in similar programs to maintain the technological edge and avoid falling behind.

President Johnson continued to pursue the matter vigorously, engaging in direct appeals to the Soviets in the spring of 1968.[121] Kosygin agreed to negotiations on July 25.[122] But what had changed? The United States could now claim to have an ABM under deployment. More importantly, the Soviets were approaching the critical threshold of parity with the United States in terms of offensive forces. The precondition seemed to be a necessity for the Kremlin to accept negotiations. Now, the United States had the added concern of maintaining parity and avoiding a Soviet thrust in strategic offensive forces that could seemingly provide them with a measure of superiority. Coupled with a credible ABM system, this advantage could undermine the

U.S. deterrent posture and erode the assured destruction capability of American strategic forces.[123]

Once again, unfortunately for the Johnson administration, history would intervene and undermine the possibility of negotiations. On August 20-21, 1968, Soviet and Warsaw Pact forces entered Czechoslovakia to quell demonstrations. Only a day earlier, on August 19, Ambassador Dobrynin had delivered National Security Advisor Rostow a note welcoming President Johnson to visit Leningrad in early October. Early on August 20, the White House received an official letter from Premier Kosygin to the president offering to open SALT talks in Geneva on September 30.[124] Rostow and Secretary of State Rusk were going to publicly announce the Leningrad summit and the opening of talks in Geneva on the morning of August 21. Eastern Bloc politics intervened, however, when Ambassador Dobrynin informed President Johnson and Rostow of the Soviet incursion at the White House at approximately 8:15 p.m. on August 20. The National Security Council met later on in the evening and agreed to delay the summit announcement indefinitely.

Even after the election of Richard Nixon, Johnson and his advisors considered the possibilities of engaging in negotiations with the Soviets, and perhaps even attending a meeting in Moscow. Johnson considered asking Nixon (or some designee of the president-elect) to join him in such an endeavor. It was generally felt among Johnson's advisors that such a diplomatic initiative would provide a unified front to the Soviets and reinforce the notion that American interest in cooperation would not end with the inauguration of a Republican president.[125] As National Security Advisor Rostow wrote to Johnson on November 14, "Moscow is clearly ready to go—and eager—if you can work it out."[126] President-elect Nixon and his principal adviser on foreign affairs, Henry Kissinger were understandably reluctant to engage in such a process before taking office and making their own assessments of the likelihood for cooperation with the Soviets and thus SALT faded for a time during the transition of power.[127]

SALT I: The ABM Treaty and an (Insufficient) Interim Agreement

President Nixon thus entered the White House with a significant amount of groundwork already prepared by the outgoing Johnson administration. However, Nixon and Kissinger sought to undertake their own assessment of the existing strategic situation and devise their own approach to arms limitation with the Soviets, and thus there was a period of relative inactivity with regard to arms limitation. For their part, the Soviet leadership maintained their willingness to negotiate in official correspondence to the incoming American president.[128] Importantly for the Nixon administration, as would become apparent, time was not necessarily on the side of the United States. The strategic balance had undergone a significant shift since the initial Johnson overtures for negotiations in 1967. The Soviets had undertaken and, to a larger degree, accomplished the goal of achieving parity with the United States in terms of offensive weapons. Thus, it was also promising that the Soviets were explicit in offering to engage in negotiations on both offensive and defensive strategic systems.[129]

This was an important signal. The Soviets had initially sought to include both offensive and defensive systems in talks with the United States during the mid-1960s. However, at that time, when the United States was perceived to have a fairly significant quantitative advantage in offensive weapons, the Johnson administration had seem preoccupied by ABM limitation without a similar regard for cooperative restraints on offensive strategic systems. With a shifting quantitative balance of strategic offensive weapons favoring the Soviets, Nixon and Kissinger were concerned at the outset with the need for some constraint of Soviet expansion to maintain relative parity and prevent a significant widening of an offensive deficit in favor of the Soviets, particularly when the United States was engaged in any building programs.[130]

Having been persistently seeking arms control talks since the invasion of Czechoslovakia in 1968, there was considerable agreement within the U.S. defense and intelligence community that the Soviets were indeed serious about cooperation. While the Nixon administration undertook internal assessments of U.S. capabilities and devised its overall strategy of "sufficiency," the Soviets continued to signal a willingness to engage in constructive negotiations. A Special National Security Estimate was delivered to the president on November 16, 1970, after the first SALT meeting had covered a wide variety of strategic issues without any concrete proposals. In effect, the first meeting was perceived to be a "feeling out" session by the Soviets to see if the new American administration was indeed serious and importantly, whether the Nixon administration actually viewed the Soviets as equal negotiating partners. This in effect would implicitly recognize Soviet parity and fulfill Nixon's public posturing relative to "sufficiency." The clear willingness of the Soviets to meet again in Vienna indicates that the Kremlin interpreted the U.S. overtures as serious and reflecting a willingness to engage in constructive talks.[131]

Thus, from early 1970 the Nixon administration had a relatively firm view that the Soviets were indeed "serious" about engaging in meaningful talks about both strategic defensive and offensive weapons, with the objective of finding a mutually beneficial arms control agreement. This notion was supported by the development of the famous "backchannel" between National Security Advisor Kissinger and Soviet Ambassador Anatoly Dobrynin in early 1970. While the negotiating teams would continue to meet formally and engage in negotiations over various details of proposals in a structured official setting, Kissinger and Dobrynin would have the ability to explore potential solutions informally and confidentially. This was perceived to provide a means of minimizing the influence of domestic political considerations and bureaucratic pressures, thus insulating the lines of communication and to maximizing the flexibility and responsiveness to the process of evaluating various issues by providing a direct link between the U.S. president and the Soviet chairman.[132]

The United States was greatly concerned, however, with Soviet offensive buildup and could not accept a treaty that focused only on strategic defenses. It was essential from the point of view of the Nixon administration to craft an agreement that would, at the very least, slow down the quantitative expansion of Soviet strategic forces, specifically its land-based ICBMs (most importantly the SS-9) and the growing Soviet submarine fleet. Strategic sufficiency was a workable policy in the face of a relatively stable opponent, but the significant growth of Soviet strategic forces, which seemed to have no end in sight, threatened to open a wide quantitative advantage in favor of the Soviets. Sufficiency could not be sustained in this context. After several months of engaging in detailed negotiations, the Soviet Union made an official entreaty to the Nixon administration. It focused on the prospects of signing an ABM agreement.[133]

Thus, the Soviets hoped to conclude, in relatively short order, an ABM agreement, and perhaps some additional codification of measures that would make accidental war less likely. Brezhnev and the Soviet leadership clearly viewed the signing of an agreement as an acknowledgment of Soviet strategic and political parity and the recognition of its power in the world. Thus, such an agreement would serve Soviet political and diplomatic interests. However, the Soviets emphasized that such an agreement would only be a first step, a beginning, to a longer process of negotiation on the control, and ultimately, the reduction of strategic forces between the superpowers. As discussed, this was not an approach the United States could accept.[134] The Nixon administration could not engage in cooperation that could potentially leave the United States worse off by failing to address the Soviet buildup of offensive weapons. Despite U.S. plans to "MIRV" Minuteman III and Poseidon missiles the seemingly unrelenting quantitative expansion of Soviet systems seemed threatening. Moreover, the U.S. ability to compete was perceived as severely constrained.[135]

The Soviets had effectively achieved parity and thus U.S. planners saw any attempt to move beyond that as an attempt to gain superiority. Thus some measure of offensive limitations was necessary. Finding a cooperative means to achieve this goal would ultimately dominate the SALT talks. Nonetheless, the seeming underlying agreement between the United States and the Soviets on the issue of ABM restrictions was promising.[136]

Opening Moves in Helsinki

The official SALT negotiations began in Helsinki with a fairly general and exploratory session that lasted from November to December 1969.[137] The delegations focused on explaining and reaching common understanding on the respective views of the superpowers relative to deterrence, nuclear war, and the roles of various strategic forces, as well as questions of verification. While little progress toward a negotiated settlement was made, the talks allowed for the exploration of issues that would prove most amendable to agreement as well as those most likely to create future impasses. With regard to the former, the Soviets indicated a genuine willingness to engage in significant limitations on ABM defenses, even perhaps a complete ban on such systems. The Soviets also expressed a willingness to discuss measures to avoid the problem of "accidental war" or the possibility of a third country initiating a superpower conflict, which had not necessarily been an issue of pressing concern for the U.S. delegation but would later result in two agreements signed in 1971. In the latter category, however, it was clear that the Soviets and the United States were far apart on the attractiveness of possible offensive limitations, particularly of Soviet ICBM and SLBM programs. Complicating the issue was the Soviet concern about U.S forward-based systems or FBS, including fighter-bombers based in Western Europe and on U.S. Navy aircraft carriers.[138] These systems provided the U.S. and NATO forces with a significant ability to hit targets within the Soviet Union, and thus should be appropriately considered, in the view of the Soviet delegation, in any discussion of offensive limitations. The issue of FBS would return to become a major source of difficulty in the third negotiating session, but the "feeling out process" of the first SALT meeting was generally viewed as a positive experience in the United States and an affirmation of the Soviet commitment to engage in a serious process. Ultimately, the Soviets would make the major concession of removing FBS from the SALT agenda in order to facilitate agreement on strategic weapons.[139]

What had become increasingly clear was that the Soviet Union was willing to engage in an ABM limitation agreement separately from a more comprehensive agreement on offensive forces. This had in essence, reflected a change in Soviet views from the Johnson administration when the Soviets were reluctant and ultimately unwilling to address ABM defenses in the absence of an agreement to limit strategic offensive programs.[140] The United States was greatly concerned, however, with Soviet offensive buildup and could not accept a treaty that focused only on strategic defenses. It was essential from the point of view of the Nixon administration to craft an agreement that would, at the very least, slow down the quantitative expansion of Soviet strategic forces, specifically its land-based ICBMs (most importantly the SS-9) and the growing Soviet submarine fleet. Strategic sufficiency was a workable policy in the face of a relatively stable opponent, but the significant growth of Soviet strategic forces, which seemed to have no end in sight, threatened to open a wide quantitative advantage in favor of the Soviets. Sufficiency could not be sustained in this context. After several months of engaging in detailed negotiations, the Soviet Union made an official entreaty to the Nixon administration. It focused on the prospects of signing an ABM agreement.[141]

Taking on MIRVs in Vienna

The second session held in Vienna, unexpectedly lasted from April to August 1970.[142] It is perhaps most important in that it was the beginning and end of any discussion of a limitation on MIRVs which had surprisingly gone unmentioned in the first session.[143] In preparation for the meetings, the United States embarked on a comprehensive government-wide evaluation of strategic programs and the perceived benefits and drawbacks of various limitation packages. Ultimately, the Nixon administration, under Kissinger's direction established a Verification Panel as a select committee within the National Security Council to coordinate planning for SALT. The result was a NSDM-49, which delineated four options for the U.S. delegation to pursue.[144] The first would be a limited agreement, which would allow both states to maintain their existing ABM systems (for the United States, this was at the level of Safeguard) with no limitation in MIRVs. A second, "comprehensive I," would include an ABM ban or limit ABM to one site to protect each superpower's capital or national command authority (NCA) with no limits on MIRVs. The third, "comprehensive II" would provide a similar ABM zero or NCA option with a ban on MIRVs, and the last option, "reductions," would make major cuts in ICBM launchers, but MIRVs deployment would be allowed. In fleshing out these proposals, the United States viewed a ceiling of 1,710 (1,054 ICBMs and 656 SLBMs) as a target for aggregate delivery vehicles, and considered freezing other systems, including manned-strategic bombers and intermediate range bombers at existing levels. While this seemed to provide a logical reference point, the aggregate ceiling represented the number of strategic offensive forces deployed by the United States for some time. For the Soviets, accepting such a limitation would likely mean completing only half of the forces under construction. During the upcoming session, the U.S. delegation was to focus on the "comprehensive II" package, while also offering the "reductions" package, which had been defined to include the removal of 100 ICBMs per year for seven years.

The proposed MIRV ban included in the "comprehensive II" package, offered and explored at Vienna may seem surprising, given the important place of MIRVs to the U.S. strategic arsenal. However, if ABM systems could indeed be negotiated down to the zero level (a complete ban) or perhaps an NCA level, then much of the initial rationale for the deployment of MIRVs would be removed. As discussed above, the initial strategic logic behind MIRV development and deployment was to maintain the deterrent capability of US offensive forces by guaranteeing the penetration of potentially large-scale Soviet ABM deployments. In the absence of such strategic defenses, MIRVs would seemingly become less important. Thus any MIRV ban necessarily assumed an ABM ban. The civilian and uniformed military leadership of the defense department was "adamantly" against a MIRV ban, but it was explored in the context of an ABM ban and with the recognition that Soviet MIRVed ICBMs would threaten the U.S. ICBM force in the near future.[145] Moreover, MIRVed Minuteman III and Poseidon systems would not come online until 1974. This lull in U.S. development played a major role in the perceived need to limit Soviet quantitative expansion in offensive forces. Nevertheless, while the MIRV ban offer was a dramatic attempt to jumpstart negotiations in a positive direction, the high threshold set out for compliance and verification necessary for agreement was viewed as something the Soviets were unlikely to accept.[146]

In return for an agreement to ban MIRVs, the United States required rigorous on-site inspection, including access to Soviet surveillance and anti-aircraft sites to ensure that these assets would not contribute to a large-scale ABM deployment.[147] Given the decisive nature of MIRV technology and the potential impact of MIRV on the strategic capabilities of a superpower that succeeded in deploying it, the United States could not accept anything less than an intrusive inspection regime. To forego MIRV and later find oneself faced with an adversary that had successfully deployed them would be a potentially catastrophic strategic failure. Moreover, the United States enjoyed a substantial lead in this technology, and its deployment would make a dramatic contribution to U.S. strategic offensive capabilities irrespective of what the Soviets would deploy. Thus, the prospect of Soviet MIRV deployment was truly too frightening to allow for any cooperative outcome short of a perfect guarantee, and too tempting to forego in the face of a dangerous adversary. As we saw in the case of the Baruch Plan, the perfect proved to be the enemy of the good and no effective MIRV ban could be achieved under SALT. Later attempts at Vienna only further reflected this fact.[148] The United States made an offer to ban the deployment of MIRVs, but would allow for the production and stockpiling of them. Given the U.S. lead in research and development, this was unpalatable to the Soviets, who instead sought to ban production and deployment, but they would effectively remain free to test, and thus catch up to the Untied States in terms of technology. These fairly obvious attempts to tailor "limitations" to the respective positions of the two superpowers with respect to MIRV development reflect both the expected value that they place upon this new application and the unwillingness to engage in any cooperation that threatened to leave one party vulnerable to these assets in the future.[149] By the middle of June, both the United States and the Soviets had moved to evaluate proposals that did not include MIRV limitations, and thus the opportunity had quickly passed.[150]

Working to Find the Right Mix

The Vienna meeting continued on, without much progress in any of the other substantive areas, save ABM. Reductions in offensive forces, particularly land-based ICBMs were met with little interest from the Soviets. Conversely, FBS continued to be a major concern of the Soviet delegation while any discussion of reduction was strenuously resisted by its American counterparts. "In retrospect, it looks as if the U.S.S.R. wanted an agreement that would stop the U.S. ABM program, with its potential for triggering a full-scale ABM competition, and one that would register American acceptance of strategic parity."[151] While constructive talks again took place concerning the accidental war and third country problems, meaningful progress on strategic limitations remained elusive. Aside from an ABM limitation the Soviets were unwilling to countenance any real reductions in offensive forces. Ultimately, the U.S. proposal presented on August 4, 1970, was a reflection of the negotiations. It included a choice of a total ABM ban or limitation to NCA, coupled with ceilings on offensive launchers and a subceiling on larger missiles. This would prove to be the basis of the final SALT I Treaty, and incorporated a fairly realistic recognition of what seemed to be possible in negotiations after almost four straight months of talks.

However, when the delegations reconvened in Helsinki from November to December 1970, the talks descended into recrimination concerning FBS, and continuing Soviet refusal to entertain the notion of strategic offensive reductions.[152] The major occurrence of this round was the Soviet offer of an "ABM only" proposal that would limit both superpowers to the defense of their capital (the NCA option). While the Soviet delegation remained committed to the continuation of talks on offensive limitation, this seemed to reflect a desire to de-link the defensive and offensive questions and achieve an agreement with the former as soon as possible. This was not acceptable to the United States, who pushed to continue discussion toward limitations on both defensive and offensive systems. During early 1971, the latter concern was heightened after the Soviets resumed building new silos for their ICBM force after a brief hiatus. With increasing popular pressure in the United States for an ABM agreement, even if separate from a comprehensive treaty, the administration shifted its views on strategic defenses. If an agreement was to be achieved, the United States would require the possession of some level of ABM to protect its land-based ICBMs against the growing Soviet offensive threat.

This would lead to a seemingly convoluted approach to negotiations during the SALT IV session held in Vienna from March until May 1971.[153] The emphasis on protection of Minuteman reflected an increasing concern on the part of the Pentagon as to the Soviet threat to the land-based ICBM force, and while such logic seemed to ignore or at least overlook the continuing viability of the other two legs of the strategic triad, the shift had another important effect. This line of discussion forced Soviet offensive weapons back into the equation. The Nixon administration was concerned that its leverage, Soviet fear of a major U.S. ABM deployment based on Safeguard, was being undermined as negotiations seemed to be pushed toward an NCA agreement or total ban separate from offensive limitations. Along with a complete ban, and capital defense, the United States now offered a "four to one" option in which the United States would maintain four ABM sites to protect the Minuteman force. The Soviets, naturally, saw this as unfair and also expressed concerns about such a site, while ostensibly protecting only the U.S. land-based deterrent, being improved to provide an area defense over time. Still, while the Soviets had intimated the possibility of some receptiveness to a missile ceiling, little progress was made. Thus, after almost two years of negotiations, the SALT process had little to show for the time and effort that had been invested.

The seeming impasse was only broken at the May 20, 1971, accord in which the Soviets agreed to incorporate an offensive agreement with an ABM limitation.[154] However, the resulting agreement would be far from the comprehensive document that U.S. leaders had initially envisioned.[155] The May agreement seemed to inject momentum back into the process, a major problem created by the Kissinger-Dobrynin backchannel was the failure to accommodate for limitations of Soviet submarines and SLBMs. Along with Soviet land-based ICBMs, SLBMs were considered a central strategic program and despite their perceived primary role as a retaliatory weapon, the sheer quantitative advantage that would be gained by the Soviet in this class of weapon in the absence of an agreed limitation was troubling. It seems that Kissinger had been too casual in response to a Dobrynin question relative to the importance of the inclusion of SLBMs in the ultimate agreement. By responding "either way," Kissinger would create a major challenge that would have to be overcome.[156] It did remove the problem of FBS, but at the expense of a comprehensive offensive agreement. Thus the May agreement was a politically useful document, but aside from "agreeing to agree" on arms control, few details were sorted out.[157]

Thus when talks resumed at Helsinki (July–September 1971) major questions remained unanswered.[158] From now on, the United States and Soviet Union were negotiating two agreements, not one comprehensive one. "Now we were to negotiate a treaty limiting ABMs together with companion and short-term constraints on strategic missile launchers."[159] The Helsinki talks did provide a useful and constructive discussion that would provide a foundation for an ABM agreement. The Soviets pressed for equality in terms of sites and the delegations worked through details on issues relative to radars and disparities in interceptors and coverage that would likely exist between the eventual U.S. and Soviet systems. This would lead to the production of a Joint Draft Text of an ABM agreement, "the first tangible results of some twenty-two months of SALT negotiations."[160] The same level of progress was not seen relative to the offensive freeze agreement. At Helsinki, the Soviets had signaled a willingness to discuss halting ICBM building, including the expansion of SS-9, which had been granted its own classification: modern large ballistic missile (MLBM). However, the Soviets were adamant that submarines, having been excluded from the May 20 accord, were not to be limited. While submarines and SLBMs were generally viewed as less threatening by the United States because of their perceived utility for second, rather than first-strike, the Soviet building program had continued throughout the entire SALT process and the sheer number of Soviet submarines and missiles, in the absence of agreed restraints, was likely to create a major imbalance in forces in the Soviets' favor. The Soviets seemingly would not budge.

Nonetheless, on October 12, 1971, President Nixon announced that he would visit the Soviet Union in May 1972. The last formal session, held in Vienna from November 1971 to January 1972 focused on fleshing out the nature of the offensive freeze in parallel with the ABM agreement.[161] SLBM remained beyond negotiation, and when pressed, the Soviets argued that it would reopen the FBS issue. Discussions of modernization and replacement of existing offensive forces were undertaken, with little tangible result other than an agreement to freeze heavy and lighter missiles, though the Soviets refused to agree to or offer a definition of a "heavy" missile. This important shift was reflected in a note from Brezhnev to Nixon, dated January 17, 1972. The Soviet leader affirmed his willingness to include an interim agreement on offensive weapons that would be included in the ultimate treaty along with a more permanent limitation on ABM defenses.[162]

Despite Brezhnev's intervention, the final Helsinki round (March–April 1972) failed to bridge the divide that had existed.[163] Progress had been made on the ABM issue, and because the agreement would be permanent, the Soviets argued that it should be equal in terms of the strategic defense capabilities that each side would be allowed to possess after its signing and entry into force. The United States had gradually moved from its "four-to-one" concept down to three and then two defense installations, but with the mission of protecting the U.S. Minuteman fields. The conclusion of ABM would move to "two and two" with both nations accepting a NCA defense and one for the defense of land-based ICBMs. Unfortunately, the SLBM question remained unanswered and the prospects were not optimistic.[164]

This discussion reflects one of the trends that had emerged in Soviet thinking during the negotiations from the "ABM only" period in the late 1970s. The Soviets increasingly spoke about the likelihood of serious follow-on talks that could develop a more comprehensive limitation of offensive weapons. Since the Soviets had been willing to move FBS and U.S. bombers off the agenda for future discussions, both the delegation and Kissinger increasingly faced pressure to defer tougher issues to subsequent negotiations. The SLBM issue was one such agenda item. However, the seemingly unfettered nature of the Soviet submarine program made U.S. acquiescence virtually impossible. In the absence of any limitations, the Soviet's deployment could dramatically alter the balance in their favor at a time when the United States was not building and would not be building until the mid-1970s. Moreover, the only perceived offsetting defensive advantages that the United States may receive from the bargaining would likely be a greater number of interceptors and thus ICBMs protected, which was little real consolation in the face of the Soviet submarine expansion. The Soviets were perceived to be firmly committed to the SLBM program.[165]

Kissinger's Secret Trip to Moscow

Thus, to attempt to break the stalemate, Kissinger arranged to take a secret advance trip to Moscow.[166] He met directly with Brezhnev on April 22, 1972.[167] Brezhnev proved to be more forthcoming on the issues than the Soviet delegation. He essentially accepted the U.S. proposal for an ABM Treaty, resolving the first major issues still on the table.[168] The Soviet leadership next presented a draft of the Interim Agreement that outlines the Soviet "compromise" on submarines and SLBMs. The Soviets had supported their demands for larger numbers of submarines (and thus SLBMs) to offset the advantages of geographically dispersed forward bases for the American submarine force, thus providing it greater access to Soviet waters. The Soviets, conversely, had no such access to forward submarine bases. Still, the Soviet proposal, while presenting a breakthrough in the negotiations, nonetheless included a significant quantitative advantage under the period of the Interim Agreement, which was tentatively scheduled for five years.[169]

The Brezhnev-Kissinger agreement would effectively break the deadlock on the SLBM issue and pave the way for the signing of the SALT I accord in Moscow during President Nixon's May 1972 Summit trip.[170] There would remain, often painstaking, negotiation over the details related to the counting of older model submarines as well as older ICBMs that would be replaced and counted toward the 62/950 ceilings. Ultimately, the consensus within the U.S. delegation and the White House was that the older assets made little real difference in the overall strategic balance and relented to the Soviet calculations, despite questions as to the assumptions of the Soviet numbers of boats to missiles. Previous U.S. assessments of the Soviet subforces and their modernization efforts did not match what the Soviet seemed to claim, but given the five-year window of the Interim Agreement, the Nixon administration had confidence that efforts could be made to offset Soviet expansion if necessary. Specifically, existing U.S. submarines scheduled for replacement could, if necessary, be retrofitted with Poseidon missiles in concert with the Trident submarines that would be deployed at the end of the period covered by the agreement to rectify any major imbalance.[171]

Success at Moscow

When President Nixon arrived in Moscow on May 23, he and Kissinger completed the diplomatic wrangling over the remaining open issues.[172] Various attempts had been made to limit the size of land-based ICBMs that could be deployed during the period of the freeze, however, they often clashed with the previously accepted consensus that modernization and qualitative improvements would not be restrained. Because of the U.S. commitment to MIRV in the replacement of Minuteman II with Minuteman III, the United States had to be careful in attempting to impose limitation directed at Soviet heavy missiles. The Soviet counter that the more important issue was the agreement to freeze the building of new silos or soft-pad missile launchers, and that increased size or volume of replacement missiles was less significant. Two other issues that received significant attention but were ultimately resolved were the location of a second Soviet ABM site and the accounting of older Soviet submarines under the numerical limits of the Interim Agreement.[173]

While far from being the comprehensive limitation that the United States had optimistically sought at the beginning of the SALT process, the Interim Agreement was nonetheless a significant agreement. From the perspective of the United States, the agreement served an important purpose of limiting the perceived acceleration of the expansion of Soviet offensive systems that threatened to seriously undermine the strategic balance between the superpowers. For Brezhnev and the Soviet regime, the freeze established a strategic balance favorable to the Soviets. More importantly, the agreement effectively acknowledged the growth of Soviet power and represented an acknowledgment of the achievement of parity with the United States.

The ABM Treaty was a comprehensive diplomatic agreement that effectively addressed the variety of highly complex technical issues associated with the ballistic missile defense systems and the differing priorities and perceived strategic needs of the two powers relative to those systems.[174] The contribution of the SALT delegations is reflected in a treaty that necessarily had to address the roles of component systems like large phased array radars, smaller targeting radars, and interceptor vehicles, as well as highly relevant issues like further research, development, and testing activities. While the subsequent 1974 agreement to decrease the number of ABM sites allowed to each superpower to one implies that a more restrictive ban may have been achievable in 1972, this misses an important point.

The two years of SALT negotiations alternating between Helsinki and Vienna clearly illustrate the difficulty of the challenge of forging a meaningful arms control agreement. Contrary to the argument that SALT I was merely a reflection of improved superpower relations associated with détente, the fact that the negotiations continued despite political crises in the Middle East, Vietnam, and the subcontinent testify to the commitment of both superpowers to forging an arms control agreement.[175] In the face of major political challenges that could have seriously undermined constructive superpower talks at various points, the SALT negotiations persisted.[176] Moreover, even after the reciprocal signals of commitment that seemed to imply a possibility of constructive arms control discussions, the hard bargaining that defined the SALT process only reaffirms the underlying fears and concerns of the two superpowers. While being victimized by defection was a major impediment to a quick agreement, clearly both states were also preoccupied with the construction of an agreement that would maintain their security over time. Thus, military expectations played a major role shaping the potential outcomes that comprised the final agreement.

CONCLUSIONS

The SALT I accords were perceived as a success because they allowed the United States and the Soviet Union to avoid a potential destabilizing competition in strategic defenses and also maintained a sense of relative balance in offensive weapons. This allowed for an enhancement in both arms race and crisis stability, even if it did not significantly decrease the probability of war, which seemed relatively low during the period. The expectations relative to the infeasibility of ABM and the maturation of delivery vehicles, along with the confidence provided by national technical means of surveillance to detect any cheating before a breakout could occur, allowed the superpowers to forge agreements as METs second and third general propositions and its specific cooperative hypotheses predict.

The most obvious perceived "failure" of the SALT I Treaty was the inability of the negotiators to forge a cooperative limitation to MIRV technology. By the end of the decade, both the United States and the Soviet Union would dramatically expand the number of warheads in the combined superpower arsenals. This "missed opportunity" is often blamed on the United States, and the unwillingness of the Nixon administration to present a serious ban on MIRVs that the Soviet Union could accept. However, as discussed above, the commitment to MIRV existed long before the SALT negotiations were initiated. As this study has argued, new applications of changing technologies that are deemed to be decisive will be prized by military and political leaders, and MIRV is an excellent example of just such an application. It presented a singularly effective capability to the potential for a Soviet breakthrough in ABM defenses. Moreover, it was a cost-effective means by which to maintain U.S. security without engaging in a quantitative expansion of U.S. strategic offensive forces that could have further spurred Soviet countermeasures, whether in the field of strategic defenses or further offensive expansion. The critical decisions to develop, produce, and deploy MIRVs on a new generation of U.S. missiles were taken long before the potential successful cooperative outcomes of SALT had ever been considered. They had thus logically assumed a central role in the planning of U.S. decision makers from the mid-1960s forward. It was virtually impossible for the United States to forego these weapons at the time of the SALT negotiations. Moreover, the MIRV-ban options that actually were offered by the United States at the Vienna talks in 1970 (and the demand for intrusive inspections) reflect the decisiveness with which MIRVs were viewed throughout the U.S. defense establishment.

Yet there was a self-defeating aspect to U.S. MIRV deployment. In retrospect, it may have indeed been in the interest of the United States and the Soviets to forego the deployment of this technology. Certainly the increasingly acute concern about the utilization of MIRV technology with the massive SS-9 delivery vehicle implied that a world without MIRVs may have actually removed such a threat to the U.S. land-based deterrent. But this argument assumes that the genie of technology could have been kept in the bottle indefinitely. Moreover, it obscures the fact that during its initial phases of development, the United States believed that it was fairly far ahead of the Soviets in the development of this technology. Regardless, as this study has argued, once a truly decisive system is discovered, there is an intense pressure to develop and deploy that system first, because it is the only real way to avoid being victimized by a concurrent or subsequent breakthrough by an enemy. It may have been possible, in retrospect, to develop MIRV technology to some level of maturity to provide policymakers with the confidence that MIRV could be rapidly deployed in the event of evidence of a Soviet breakthrough, but given the perceived increase in capabilities that MIRV could provide, such an approach would have been exceedingly difficult. Finally, the criticism of the failure to ban MIRVs implies that the Soviets would have truly desired such an outcome. This assumption is also tenuous at best. As discussed above, states will be extremely wary of locking themselves into a position of perceived technological inferiority relative to military capabilities. This wariness is only intensified with the decisiveness of the weapon in question. If the United States was willing to agree to forego the deployment of MIRV, and if the Soviets were allowed to develop MIRV technology, but similarly not deploy, and if the requisite inspection and compliance requirements could support such a bargain, then MIRV may have been limited. Such a series of events seems highly unlikely, if not impossible, given the nature of the superpower rivalry and the underlying dynamic of the arms race.

The criticism focused on the failure to ban MIRV also reflects a more general weakness of SALT that is perhaps overstated: the perceived limited scope of the ultimate agreements. This misses a fairly straightforward but important counterargument: SALT I was only viewed as a first step toward greater, more comprehensive superpower arms control agreements. When one considers the strategic asymmetries that existed between the United States and the Soviet Union, and the differing priorities and subjective assessments of security engaged in any discussion of strategic arms control, the limited nature of SALT I should not be judged to its detriment. To the contrary, the Soviet concern for NATO assets, as well as U.S. strategic bombers, and most importantly, forward-based systems that could easily deliver nuclear warheads to the sovereign territory of the Soviet Union threatened to derail SALT before any meaningful agreement could be reached. "Comprehensiveness" can often be a hindrance to arms control negotiations. As the subsequent history of the SALT process would reflect, significant progress was made toward addressing many of the questions of offensive strategic force levels, bomber forces, and FBS in the SALT II process culminating in Vladivostok accord.[177] Though SALT II was never ratified by the United States, it served as a de facto arms control agreement well into the 1980s, providing relatively stable expectations for planners on both sides, despite a significant deterioration in the political relationship between the superpowers.

Ultimately, the inability to conclude an agreement on MIRV also provides strong evidence for MET's first general proposition as well as both the specific propositions, $\mathbf{H}_{\mathbf{Lead1}}$ and $\mathbf{H}_{\mathbf{Lead2}}$. Initiated in the early 1960s, MIRV was perceived as a decisive application and thus the United States had little desire to restrain it through cooperation. Similarly, it was not included in the Johnson administration approaches. It was only considered and offered to the Soviets after the United States became concerned about a potential Soviet MIRV, but by then there was little chance that Moscow would lock itself into a position of inferiority as HLag1 predicts. Thus the variation of outcomes within the SALT one case provides robust support for the general and specific propositions of MET.

Alternative hypotheses perform poorly in the SALT I cases. Offensive realism would have little predictive value. As the long history of the negotiations reflects, the eventual successful cooperation was neither easy nor predetermined. More importantly, limiting any potentially decisive weapon system (like ABM) when confronted with a persistent rival would seem antithetical to the predictions of offensive realism. Employing a notion of deterrence dominance, analogous to defense dominance, may allow defensive realism to provide some predictive utility on the ABM Treaty and the Interim Agreement, but it would seemingly also predict a successful MIRV agreement. So once again, the variation within the case underscores the limitations of defensive realist analysis. More generally, it is also less clear why the SALT I Accords could not have been successfully concluded earlier, perhaps under the Johnson administration, given the predictions of defensive realism. The timing of the cooperation is inadequately addressed in what is a relatively static theory. Given the importance of (unilateral) national technical means and the minimal role of institutions in the SALT story, the case provides little support for institutionalist arguments concerning cooperation.

Finally, in terms of domestic-level alternatives, two prevailing historical issues should be considered. First, Congressional support and funding for Sentinel ABM was tenuous during the period that the Nixon administration was negotiating SALT. This has led some scholars to argue that therefore ABM was a formal codification of domestic economic imperatives to avoid paying for an unpopular program. Given that many technological hurdles associated with effective strategic defense continue to confound researchers, and considering the perception that President Johnson initially signed off on a "thin" Safeguard deployment to signal to Moscow that the United States could compete in strategic defenses even while it preferred negotiations to avoid such a competition, this domestic argument is important. I would argue that the debates over ABM within the executive and legislative branches actually support the depiction of expectations about ABM assumed in this case and considered in the context of MET. Despite significant investments, strategic defenses seemed infeasible in any relevant time frame. However, so long as the Soviets continued to work on the problem, the military and political leadership of the United States could not simply unilaterally surrender the cause. Only with the security cooperation that limited and eventually completely prohibited ABM sites and associated infrastructure could either states move away from the challenge of achieving a breakthrough. Even if funding had been cut for Sentinel, such a move would have likely generated pushback from military leaders and the issue would have remained an open one. Such a development may have indeed undermined the ability of the United States to conclude the ABM Treaty, but it is beyond the scope of the argument to consider this further. The eventual success of the Reagan administration to secure support for the Strategic Defense Initiative would seemingly undermine the notion that ABM was widely viewed as effectively unnecessary and the United States could simply stop research and development (if not deployment) without any concurrent Soviet agreement that would end the race in strategic defenses. Second, some have argued that while the United States and Soviets obviously maintained deep political differences, Moscow had come to accept the concept of secure retaliation and Mutually Assured Destruction (MAD) as many U.S. leaders had. This ideational consensus was thus a necessary condition for the successful conclusion of the SALT I accords. While I discussed above that it was indeed difficult to ascribe "security-seeking" motives to Brezhnev and the Soviet leadership, the argument that Moscow embraced MAD simply goes too far and subsequent Soviet behavior would ultimately erode superpower relations though the SALT I accords would remain in effect, recognized by both U.S. and Soviet leaders despite changes in the security and diplomatic environments in the late 1970s and early 1980s.[178]

NOTES

1. On SALT I see: John H. Barton and Lawrence D. Weiler, eds., *International Arms Control: Issues and Agreements* (Stanford, CA: Stanford University Press, 1976), 172–208; Coit D. Blacker and Gloria Duffy, eds., *International Arms Control: Issues and Agreements*, 2nd ed. (Stanford, CA: Stanford University Press, 1984), 219–54; George Bunn, *Arms Control by Committee: Managing Negotiations with the Russians* (Stanford, CA: Stanford University Press, 1992), 106–41; Philip J. Farley, "Strategic Arms Control, 1967–1987," in *U.S.-Soviet Security Cooperation: Achievements, Failures, Lessons*, ed. Alexander L. George, Philip J. Farley, and Alexander Dallin (New York: Oxford University Press, 1988); Raymond L. Garthoff, *Detente and Confrontation: American-Soviet Relations from Nixon to Reagan* (Washington, DC: Brookings Institution, 1985); Jerome H. Kahan, *Security in the Nuclear Age: Developing U.S. Strategic Arms Policy* (Washington, DC: Brookings Institution, 1975); Roger P. Labrie, ed., *SALT Handbook: Key Documents and Issues 1972–1979* (Washington, DC: American Enterprise Institute, 1980); John Newhouse, *Cold Dawn: The Story of SALT* (New York: Holt, Rinehart and Winston, 1973); Fen Olser Hampson, "SALT I: Interim Agreement and Abm Treaty," in *Superpower Arms Control: Setting the Record Straight*, ed. Albert Carnesale and Richard N. Haass (Cambridge: Ballinger, 1987); John B. Rhinelander, "The SALT I Agreements," in *SALT: The Moscow Agreements and Beyond*, ed. Mason Willrich and John B. Rhinelander (New York: Free Press, 1974); Condoleeza Rice, "SALT and the Search for a Security Regime," in *U.S.-Soviet Security Cooperation: Achievements Failures Lessons*, ed. Alexander L. George, Philip J. Farley, and Alexander Dallin (Stanford, CA: Stanford University Press, 1988); Chalmers M. Roberts, "The Road to Moscow," in *SALT: The Moscow Agreements and Beyond*, ed. Mason Willrich and John B. Rhinelander (New York: Free Press, 1974); Mason Willrich, "SALT I: An Appraisal," in *SALT: The Moscow Agreements and Beyond*, ed. Mason Willrich and John B. Rhinelander (New York: Free Press, 1974); Mason Willrich and John B. Rhinelander, eds., *SALT: The Moscow Agreements and Beyond* (New York: Free Press, 1974).

2. Ashton B. Carter and David N. Schwartz, eds., *Ballistic Missile Defense* (Washington, DC: Brookings Institution, 1984); David N. Schwartz, "Past and Present: The Historical Legacy," in *Ballistic Missile Defense*, ed. Ashton B. Carter and David N. Schwartz (Washington, DC: Brookings Institution, 1984); Sayre Stevens, "The Soviet Bmd Program," in *Ballistic Missile Defense*, ed. Ashton B. Carter and David N. Schwartz (Washington, DC: Brookings Institution, 1984); Ernest J. Yanarella, *The Missile Defense Controversy: Strategy, Technology, and Politics, 1955–1972* (Lexington: University Press of Kentucky, 1977).

3. John M. Clearwater, *Johnson, Mcnamara, and the Birth of SALT and the ABM Treaty 1963–1969* (Dissertation.com, 1996); Alain C. Enthoven and K. Wayne Smith, *How Much Is Enough? Shaping the Defense Program 1961–1969* (Santa Monica: RAND Corporation, 2005); Robert S. McNamara, *The Essence of Security* (New York: Harper and Row, 1969).

4. Desmond Ball, *Politics and Force Levels: The Strategic Missile Program of the Kennedy Administration* (Berkeley: University of California Press, 1980), 88–106; Edgar M. Bottome, *The Missile Gap: A Study of the Formulation of Military and Political Policy* (Cranbury, NJ: Fairleigh Dickinson University Press, 1971); McGeorge Bundy, *Danger and Survival: Choices about the Bomb in the First Fifty Years* (New York: Random House, 1988), 319–57.

5. Kahan, *Security in the Nuclear Age*; Lawrence J. Korb, *The Fall and Rise of the Pentagon: American Defense Policies in the 1970s* (Westport, CT: Greenwood Press, 1979).

6. Henry A. Kissinger, *American Foreign Policy, Three Essays* (New York: W. W. Norton, 1969), 65–78.

7. Georgi Arbatov, *The System: An Insider's Life in Soviet Politics* (New York: Times Books, 1992), 175.

8. Garthoff, *Detente and Confrontation*, 199–247.

9. Robert Dalleck, *Nixon and Kissinger: Partners in Power* (New York: HarperCollins, 2007); Seymour M. Hersh, *The Price of Power: Kissinger in the Nixon White House* (New York: Summit Books, 1983); Walter Isaacson, *Kissinger: A Biography* (New York: Simon & Schuster, 2005); Marvin Kalb and Bernard Kalb, *Kissinger* (Boston: Little, Brown, 1974).

10. Ibid., 132–33.

11. Newhouse, *Cold Dawn: The Story of SALT*, 166–78.

12. Garthoff, "Mutual Deterrence, Parity and Strategic Arms Limitation in Soviet Policy," 104–6.

13. Kahan, *Security in the Nuclear Age*, 163.

14. Schwartz, "Past and Present: The Historical Legacy," 331–33; Yanarella, *The Missile Defense Controvery*, 26–42.

15. On technological challenges, see: Weiner, "Systems and Technology."

16. Greenwood, *Making of the MIRV*, 54; Tammen, *MIRV and the Arms Race*, 81–90; Lapp, *Arms beyond Doubt*.

17. Seen for example, "Memorandum from the Secretary of Defense (McNamara) to the President, RE: Production and Deployment of Nike-X," January 17, 1967, in *FRUS, 1964–1969*, XI, 421–23; Kahan, *Security in the Nuclear Age*, 99–101; Yanarella, *The Missile Defense Controversy*, 60–119.

18. See, for example, "Memorandum from JCS (Wheeler) to Secretary of Defense (McNamara)," January 19, 1967, in *FRUS, 1964–1968, XI*, 426–29.

19. Secretary McNamara admitted as much to Canadian Minister of Defense Hellyer in a meeting on September 12, 1967. See Memorandum of Conversation in *Foreign Relations of the United States, 1964–1968, National Security Policy*, vol. 10 (Washington, DC: U.S. Department of State), 607.

20. *FRUS, 1964–1968, XI*, 423.

21. Ibid.

22. Ibid.

23. *FRUS, 1964–1968,* X, 563.

24. Clearwater, *Johnson, Mcnamara, and the Birth of Salt and the Abm Treaty 1963–1969*, 232–33; *Foreign Relations of the United States, 1969–1976, the Soviet Union*, vol. 12 (Washington, DC: U.S. Department of State), 512–31; Lyndon Baines Johnson, *The Vantage Point: Perspectives of the Presidency 1963–1969* (New York: Holt, Rinehart and Winston, 1971), 481–85; Newhouse, *Cold Dawn*, 94–96.

25. On the aborted announcement and the Czech crisis, see *FRUS, 1964–1968, XI*, 689–90.

26. House of Representatives, Committee on Armed Services, *Hearings on Military Posture*, 91st Cong., 2nd Sess., 1970, 7110.

27. Kahan, *Security in the Nuclear Age*, 151–53; Newhouse, *Cold Dawn: The Story of Salt*, 150–52; Yanarella, *The Missile Defense Controversy*, 143–80.

28. Committee on Armed Services, US House of Representatives, *Military Posture*, 91st Cong., 2nd Sess., February–April 1970, 6846.

29. Ibid., 7914–15.

30. On the domestic political situation see: Robert A. Bernstein and William W. Anthony, "The ABM Issue in the Senate, 1968–1970: The Importance of Ideology," *American Political Science Review* 68, no. 3 (1974); Robert J. Bresler and Robert C. Gray, "The Bargaining Chip

and Salt," *Political Science Quarterly* 92, no. 1 (1977); Alton Frye, *A Responsible Congress: The Politics of National Security* (New York: McGraw-Hill Book Company for the Council on Foreign Relations, 1975).

31. Ibid., 8325.
32. Ibid., 6842.
33. Enthoven and Smith, *How Much Is Enough* , 184–96
34. On U.S. missile developments, see York, *Race to Oblivion: A Participant's View of the Arms Race*, 75–105; Ball, *Politics and Force Levels*, 6–8.
35. Ball, *Politics and Force Levels*, 3–58; Kahan, *Security in the Nuclear Age*, 84–94.
36. Enthoven and Smith, *How Much Is Enough?*, 165–96.
37. Ball, *Politics and Force Levels*, 42, 50.
38. Collins, *American and Soviet Military Trends since the Cuban Missile Crisis*, 88–94, 96–102; Enthoven and Smith, *How Much Is Enough?*, 206–10.
39. *FRUS, 1964–1968*, XI, 413.
40. York, *Race to Oblivion*, 147–69.
41. Ball, *Politics and Force Levels*, 155.
42. York, *Race to Oblivion: A Participant's View of the Arms Race*, 160–61.
43. Ibid., 150.
44. Ibid., 153.
45. At some level of quantitative superiority, counterforce missions may become conceivable, but the United States moved away from this view under McNamara's leadership. Enthoven and Smith, *How Much Is Enough?*, 183–84.
46. Gerard C. Smith, *Doubletalk: The Story of Salt I* (Lanham, MD: University Press of America, 1985), 118.
47. York, *Race to Oblivion*, 156.
48. Ball, *Politics and Force Levels*, 155.
49. Enthoven and Smith, *How Much Is Enough?*, 67, 208; Kahan, *Security in the Nuclear Age*, 209–11.
50. *FRUS, 1964–1968*, XI, 659.
51. On the development of Corona, see Thomas Graham Jr. and Keith A. Hansen, *Spy Satellites and Other Intelligence Technologies That Changed History* (Seattle: University of Washington Press, 2007).
52. Ibid., 39.
53. "Paper Prepared by the Interagency SALT Steering Committee (Undated)," *Foreign Relations of the United States, 1969–1976*, vol. 32; *SALT I, 1969–1972*, 29.
54. "Paper Prepared in the Arms Control and Disarmament Agency (ACDA–3356) Washington, June 11, 1969. "A 'Stop Where We Are' Proposal for SALT." *FRUS 32, 1969–1972*, 44.
55. "Memorandum from Director of Central Intelligence Helms to President's Assistant for National Security Affairs (Kissinger), Washington, August 4, 1969," in *FRUS 32, 1969–1972*, 128.
56. "Paper Prepared by the Interagency SALT Steering Committee (Undated)," 35.
57. Ibid.
58. Ibid., 8170–71.
59. Enthoven and Smith, *How Much Is Enough?*, 181–83; Greenwood, *Making the Mirv*, 69.
60. Enthoven and Smith, *How Much Is Enough?*, 193.
61. "Wheeler to McNamara," *FRUS, 1964–1968*, XI, 428.
62. "Special National Intelligence Estimate, 11–13–68," July 18, 1968, in *FRUS, 1964–1968, XI*, 647.
63. "Memorandum from the JCS (Wheeler) to McNamara," January 19, 1967, in
64. Ibid., 429.
65. Ibid., 428.
66. Ibid., 659.
67. Ibid., 666–67.
68. Ibid., 647.
69. Ibid., 711.

70. *Hearings on Military Posture*, 6844.

71. Ibid., 7110.

72. On Soviet strategy and doctrine during the period, please see: Honore M. Catudal, *Soviet Nuclear Strategy from Stalin to Gorbachev: A Revolution in Soviet Military and Political Thinking* (Atlantic Highlands: Humanities Press International, Inc., 1988), John Erickson, ed., *The Military-Technical Revolution: Its Impact on Strategy and Foreign Policy* (New York: Frederich A. Praeger Publishers, for the Institute for the Study of the USSR, 1966), Holloway, *The Soviet Union and the Arms Race*, Andrei A. Kokoshin, *Soviet Strategic Thought, 1917–1991* (Cambridge: The MIT Press, 1998), Derek Leebaert, ed., *Soviet Military Thinking* vol. George Allen & Unwin (London: 1981), Michael MccGwire, *Military Objectives in Soviet Foreign Policy* (Washington, DC: The Brookings Institution, 1987), P. H. Vigor, *The Soviet View of Disarmament* (New York: St. Martin's Press, 1986), Thomas W. Wolfe, *Soviet Strategy at the Crossroads* (Cambridge: Harvard University Press, 1964).

73. Christoph Bluth, *Soviet Strategic Arms Policy before Salt* (Cambridge: Cambridge University Press, 1992), 83–112, Alexander Dallin, ed., *The Soviet Union, Arms Control, and Disarmament: A Study of Soviet Attitudes* (New York: School of International Affairs, Columbia University, 1964), 93–103, Joseph D. Douglass and Amoretta M. Hoeber, *Soviet Strategy for Nuclear War* (Stanford, CA: Hoover Institution Press, 1979), 7–33, Raymond L. Garthoff, *Soviet Strategy in the Nuclear Age* (New York: Frederick A. Praeger, 1958), 61–91.

74. Bluth, *Soviet Strategic Arms Policy before Salt*, 17, William T. Lee and Richard F. Staar, *Soviet Military Policy since World War II* (Stanford, CA: Hoover Institution Press, 1986), 61–78.

75. Bluth, *Soviet Strategic Arms Policy before Salt*, 107–20, Kahan, *Security in the Nuclear Age*, 116, MccGwire, *Military Objectives in Soviet Foreign Policy*, 13–35, Wolfe, *Soviet Strategy at the Crossroads*, 56–78.

76. There continued to be debate about whether the Soviets actually accepted "assured destruction and "war fighting" remained a concern of the military. However, it seems somewhat more appropriate to argue that deterrence became the primary objective of Soviet strategic forces. Raymond L. Garthoff, "Mutual Deterrence and Strategic Arms Limitation in Soviet Policy," *International Security* 3, no. 1 (1978), Garthoff, "Mutual Deterrence, Parity and Strategic Arms Limitation in Soviet Policy.", Sienkiewicz, "Soviet Nucelar Doctrine and the Prospects for Strategic Arms Control."

77. "Special National Intelligence Estimate, 11–16–68, 'The Soviet Approach to arms Control,'" in *FRUS, 1964–1968, XI*, 734.

78. Bluth, *Soviet Strategic Arms Policy before Salt*, 209.

79. Ibid., 202–11, Lee and Staar, *Soviet Military Policy since World War II*, 66–70.

80. Bluth, *Soviet Strategic Arms Policy before Salt*, 206-09, *The Soviet Military Technological Challenge*, (Washington, DC: The Center for Strategic Studies, Georgetown University, 1967), 60–70, Nikolai A. Talensky, "Anti-Missile Systems and Disarmament," in *The Military-Technical Revolution: It Impact on Strategy and Foreign Policy*, ed. John Erickson (New York: Frederick A. Praeger, 1966).

81. Anatoly Dobrynin, *In Confidence: Moscow's Ambassador to America's Six Cold War Presidents* (Seattle: University of Washington Press, 1995), 213.

82. Robert P. Berman and John C. Baker, *Soviet Strategic Forces: Requirements and Responses* (Washington, DC: The Brookings Institution, 1982), Stevens, "The Soviet Bmd Program," 203-04.

83. *FRUS, 1964–1968, XI*, 735.

84. Kahan, *Security in the Nuclear Age*, 112–16, *The Soviet Military Technological Challenge*, 51–56.

85. Kahan, *Security in the Nuclear Age*, 165–69.

86. MccGwire, *Military Objectives in Soviet Foreign Policy*, 237–41.

87. Bluth, *Soviet Strategic Arms Policy before Salt*, 118.

88. MccGwire, *Military Objectives in Soviet Foreign Policy*, 238.

89. Bluth, *Soviet Strategic Arms Policy before Salt*, 105, 17–18.

90. Michael MccGwire, "Soviet Naval Doctrine and Strategy," in *Soviet Military Thinking*, ed. Derek Leebaert (London: George Allen & Unvwin, 1981), *The Soviet Military Technological Challenge*, 46–49.
91. MccGwire, "Soviet Naval Doctrine and Strategy," 192–98.
92. MccGwire, *Military Objectives in Soviet Foreign Policy*, 240–41.
93. Bluth, *Soviet Strategic Arms Policy before Salt*, 198, MccGwire, *Military Objectives in Soviet Foreign Policy*, 241, MccGwire, "Soviet Naval Doctrine and Strategy," 173–75.
94. Bluth, *Soviet Strategic Arms Policy before Salt*, 192–93.
95. MccGwire, *Military Objectives in Soviet Foreign Policy*, 239.
96. Bluth, *Soviet Strategic Arms Policy before Salt*, 245.
97. Newhouse, *Cold Dawn: The Story of Salt*, 160–61.
98. Kahan, *Security in the Nuclear Age*, 165.
99. Collins, *American and Soviet Military Trends since the Cuban Missile Crisis*, 94.
100. Ibid., 212.
101. Enthoven and Smith, *How Much Is Enough?*, 184–90, Lawrence Freedman, *U. S. Intelligence and the Soviet Strategic Threat* (Princeton, NJ: Princeton University Press, 1986), 83, McNamara, *The Essence of Security*, 60–61.
102. *Foreign Relations of the United States, 1964–1968, Arms Control and Disarmament*, vol. XI (Washington, DC: U.S. Department of State), 412.
103. Ibid., 422–23.
104. Ibid., 633–35, 77.
105. Ibid., 595.
106. Ibid., 411.
107. Ibid., 417.
108. Ibid., 433.
109. Yanarella, *The Missile Defense Controversy*, 118.
110. *FRUS, 1964–1968, XI*, 428, *Foreign Relations of the United States, 1964–1968, National Security Policy*, vol. X (Washington, DC: U.S. Department of State), 446–52, Tammen, *Mirv and the Arms Race*, 100-02.
111. *FRUS, 1964–1968, XI*, 453.
112. Ibid., 406.
113. Ball, *Politics and Force Levels*, 46–53, Enthoven and Smith, *How Much Is Enough?*, 208.
114. *FRUS, 1964–1968, XI*, 453.
115. Ibid., 565–72, 677.
116. Ibid., 410.
117. Ibid., 431.
118. Ibid.
119. Johnson, *The Vantage Point*, 481–85.
120. Enthoven and Smith, *How Much Is Enough?*, 190–94, Morton H. Halperin, "The Decision to Deploy the Abm: Bureaucratic and Domestic Politics in the Johnson Administration," *World Politics* 25, no. 1 (1972); McNamara, *The Essence of Security*, 163–66.
121. *FRUS, 1964–1968, XI*, 590–92.
122. Ibid., 652.
123. Ibid., 678.
124. Ibid., 689–90.
125. Ibid., 739.
126. Ibid., 738.
127. Newhouse, *Cold Dawn*, 136–37.
128. *Foreign Relations of the United States, 1969–1976: The Soviet Union*, vol. 12 (Washington, DC: U.S. Department of State), 46.
129. Ibid.
130. Ibid., 51.
131. Ibid., 410.
132. Ibid., 434.
133. Ibid., 547.

134. Ibid., 549–50.
135. *Foreign Relations of the United States, 1969–1976: The Soviet Union*, vol. 14 (Washington, DC: U.S. Department of State), 218.
136. *FRUS, 1969–1976, XII*, 550.
137. Newhouse, *Cold Dawn*, 172–76; Smith, *Doubletalk*, 75–107.
138. .On FBS, see Kahan, *Security in the Nuclear Age*; Stephen M. Millett, "Forward-Based Nuclear Weapons and Salt I," *Political Science Quarterly* 98, no. 1 (1983): 178–80.
139. Garthoff, *Detente and Confrontation*, 158.
140. "Memorandum of Conversation: Kissinger, Dobrynin," January 10, 1970, in *FRUS, 1969–1976, XII*, 434.
141. "Aide-Memoire from the Soviet Union," July 7, 1970, in
142. Newhouse, *Cold Dawn*, 179–92; Smith, *Doubletalk*, 121–53.
143. On "The Great MIRV Mystery," see Smith, *Doubletalk*, 154–78.
144. "National Security Decision Memorandum 49" (Nixon Archives (Online), March 27, 1970); Smith, *Doubletalk*.
145. Garthoff, *Detente and Confrontation*, 134–35; Smith, *Doubletalk*, 161.
146. Smith, *Doubletalk*, 157–59.
147. Ibid., 170.
148. Newhouse, *Cold Dawn: The Story of Salt*, 179–83; Smith, *Doubletalk*, 172–78.
149. Kissinger, *White House Years*, 212.
150. Garthoff, *Detente and Confrontation*, 136–42.
151. Smith, *Doubletalk*, 124.
152. Newhouse, *Cold Dawn*, 192–98; Smith, *Doubletalk*, 179–98.
153. Newhouse, *Cold Dawn*, 198–221 Smith, *Doubletalk*, 201–21.
154. "Statement by President Nixon in the Strategic Arms Limitation Talks," May 20, 1971, *Documents on Disarmament* (Washington, DC: United States Arms Control and Disarmament Agency, 1971), 298.
155. Smith, *Doubletalk*, 225.
156. Ibid., 228–29.
157. Ibid., 245–46.
158. Newhouse, *Cold Dawn*, 225–33; Smith, *Doubletalk*, 247–79.
159. Smith, *Doubletalk*, 248.
160. Ibid., 278.
161. Newhouse, *Cold Dawn*, 234–40; Smith, *Doubletalk*, 319–50.
162. *Foreign Relations of the United States, 1969–1976: The Soviet Union*, vol. 14 (Washington, DC: U.S. Department of State), 138.
163. Smith, *Doubletalk*, 351–70.
164. *FRUS, 1969–1976, XIV*, 419.
165. Ibid., 420.
166. Smith, *Doubletalk*, 370–78.
167. Documentation relative to Kissinger's trip to Moscow and meetings with Brezhnev can be found in *FRUS, 1969–1976, XIV*, 516–37.
168. "Memorandum of Conversation" April 22, 1972, in *FRUS, 1969–1976, XIV*, 529–31.
169. Ibid., 530.
170. "Memorandum from Kissinger to President Nixon," April 24, 1972, in *FRUS, 1969–1976, XIV*, 642–43.
171. "Memorandum from Kissinger to President Nixon," April 24, 1972, in *FRUS, 1969–1976, XIV*, 625.
172. "Memorandum of Conversation, Nixon, Kissinger, Brezhnev," May 23, 1972, in *FRUS, 1969–1976, XIV*, 1007–21.
173. "Memorandum of Conversation, Nixon, Kissinger, Brezhnev," May 23, 1972, in *FRUS, 1969–1976, XIV*, 1010–21.
174. The text of the ABM Treaty and the Offensive agreement can be found in "Strategic Arms Limitation Agreements Transmitted to Congress," *Department of State Bulletin* 67, no. 1723 (1972).

175. See, for example, "Memorandum of Conversation (Alexander Haig, President Nixon)," May 5, 1972, in *FRUS, 1969–1976, XIV*, 688–89.

176. Smith acknowledges this in recalling U.S. action against Haiphong harbor and other targets in North Vietnam. Smith, *Doubletalk*, 382–83.

177. On SALT II, see Strobe Talbott, *Endgame: The Inside Story of SALT II* (New York: HarperCollins, 1979).

178. See Garthoff, *Détente and Confrontation*, especially 591–619 and 801–27.

Conclusions

This book has attempted to develop a realist theory of security cooperation that focuses on the critical importance of the influence of technological change on state behavior. Building upon existing theories, MET links the security strategies of states to their decisions to seek out or engage in cooperation with the objective of maintaining or improving their security. Specifically, qualitative arms control, which involves the prohibition, reduction, or limitation of certain classes of weapons, may offer states a means of decreasing the probability of conflict by removing those weapons deemed most threatening or useful in launching aggressive war from state arsenals. As MET explains, the willingness to engage in arms control negotiations will be shaped by the expectations developed by political and military leaders relative to the potential military value of specific weapons systems—most importantly whether they represent truly new and decisive capabilities or not. Considering other important parameters such as the security environment in which weapons are developed and the timing of innovation (whether a state is a leader or a laggard), states will prove more or less enthusiastic about the prospects of limitations of their capabilities. Even with one or more willing partners, the outcome of these negotiations are likely to be determined by the confidence leaders possess in prevailing means of surveillance to detect (and deter) cheating. Incorporating the more circumspect security-seeking motives of defensive realist models and the fears of future change emphasized by offensive realism, the theory fills an important gap in the theoretical literature on cooperation in the security realm. MET explains the conditions under which we are most likely to see successful arms control as well as the conditions where agreements are unlikely, and where states are likely to compete, in a more comprehensive way than existing offensive or defensive realist theories, as well as institutionalist approaches. The evidence presented in the case study chapters seem to confirm that MET explains both the cooperative and competitive impulses driving state decisions, as well as the outcomes of attempts to limit or proscribe various armaments.

This chapter considers the implications of MET for the broader international relations (IR) and security studies literatures. It will then briefly consider some of the practical policy-relevant applications of the logic of MET for several pressing issues in the current day world politics.

MILITARY EXPECTATIONS THEORY AND SECURITY STUDIES

Since the 1850s, and more obviously since the turn of the twentieth century, states have become active players in the development of new military applications, seeking to exploit technological change in order to improve their security and improve their probability of survival. Applications that are perceived to provide dramatic new capabilities are prized and states will endeavor to develop and deploy them as soon as possible. Introducing a decisive new weapon can significantly improve state security unilaterally. At the same time, once a state is aware of a new, decisive application it will fear that others may develop and deploy it as well, perhaps leaving the first state at a distinct and possibly catastrophic disadvantage. These dual concerns, one based on opportunity and one based on fear, will drive states to focus the necessary resources toward the acquisition of this new application to avoid being victimized by technological breakthroughs by potential competitors.

The same expectations relative to the military impact of applications of technological change that drive these decisions on acquisition and deployment will also condition the willingness of the state to engage in arms control. Where states expect military applications to be decisive, offering dramatic new capabilities far beyond existing deployed systems, they will value them highly, but also fear them in the hands of an adversary. Only under a benign security environment where a leading state already possesses these decisive weapons would we expect a state to consider arms control initiatives. However, precisely because the weapon in question is so decisive, a laggard state that does not yet possess the weapon cannot lock itself into a position of permanent vulnerability. So long as it possesses the industrial and technological capacities of a great power, it will attempt to rectify this dangerous imbalance by developing and deploying the breakthrough weapon. Even if cooperation is attempted, difficulties associated with developing a surveillance system capable of credibly safeguarding the laggard from the leading state's advantage and preventing the laggard from achieving a breakout is simply impossible. In a threatening security environment, typically where an intense security dilemma or arms race dynamic already exists, states will work hard to obtain new weapons that are expected to be decisive, expecting that an adversary's research and development activities are directed toward a similar goal. Moreover, states will have little incentive or desire to seek out or favorably respond to arms control initiatives that could hinder their progress and provide an adversary with an exploitable advantage.

Conversely, where states expect new military applications to provide capabilities similar to those offered by existing weapons, or perhaps only marginal improvements over existing capabilities, leaders will be more willing to cooperate with other states if such cooperation can maintain or improve their security. Even in a relatively uncertain or troubled security environment, the prevalence of weapons system predicated on mature technologies may provide leaders with the ability to signal benign motives, dampen crisis or arms race instabilities, or even decrease the probability of war through reassurance built upon arms control agreements. If states can further be confident in their abilities to monitor the agreements and mitigate the downside risks and costs associated with cheating, then arms control cooperation can be a useful diplomatic tool for states and tangibly improve the security of the respective actors.

These arguments are based upon the assumption that states are aware that technology is a dynamic property. Technological change is a constant process, but the implications of that change are not always linear or even. Certain periods may be defined by dramatic change in which innovative states are able to exploit technology to realize a wide variety of applications. Cooperation under these circumstances would seem to be difficult, if not impossible. During other periods, technology may prove to be relatively stable, offering little change in military capabilities, even for states that are innovative. This reflects the reality (and limitations) of innovation. Technology seems to possess both exogenous (system-level) and exogenous (state or unit-level) properties. In the latter, we can expect that states will attempt to manipulate technological change to their advantage, but the systemic component of technology will circumscribe what is possible at any given time. The combination of these two interacting flows of technological change creates one fundamental implication for states seeking to survive in an anarchic system: compete or risk catastrophe. Thus, in general, we would expect technological change to exacerbate the uncertainty inherent in the anarchic international system. However, states must develop strategies to manage this uncertainty and it is clear that technology does not change at the same rate all of the time and for long periods the military implications of technological change may be quite limited. Moreover, rational leaders cannot simply base their choices on worst-case thinking because they simply cannot afford it. All states face some level of resource constraints, and thus political and military leaders must approach the development and acquisition of military applications in a strategic way. States will seek to understand the underlying nature of existing technology and attempt to forecast how that will change in order to best adapt to potential change over time. States will also attempt to understand what other states are likely to develop, and what the potential future security environment may look like.

Where offensive realism would predict little cooperation in the security realm, and defensive realist approaches would ground their predictions in the nature of the existing security environment, MET explains the variations in state behavior that neither approach can adequately address. Consider the Baruch Plan and in the discussion of MIRV technology during the SALT I negotiations. In both cases, the United States had achieved a major breakthrough, and while its leaders perceptively understood the implications of the Soviet acquisition of these weapons, and attempted to cooperatively prohibit them, only "perfectly" effective regimes of inspection and compliance could assuage American fears. In a less acute but equally instructive way, we also saw the inability of the leading naval powers to manage the emerging challenges of naval aviation, cruisers, and submarines in what was an otherwise "successful" arms limitation agreement at Washington that decreased the probability of war between the United States and Japan in the Western Pacific and dampened a burgeoning global naval rivalry between Great Britain and the United States.

Conversely, the nature of technological change may also provide conditions where states are able to forge credible agreements to limit their capabilities in order to cooperatively improve their security. This will typically be during periods when technologies are perceived to have "matured," and the military applications are unlikely to provide decisive new capabilities, even to highly innovative states. Under such conditions, states with an existing quantity of weapons deemed sufficient to maintain national security can forge agreements to maintain those levels to thus maintain their security without threatening others. In this way, technology may provide states with opportunities to cooperatively mitigate the security dilemma. Even under these more benign conditions, military expectations of the impact of technological change can only be a necessary, but not sufficient factor in successful cooperation. States will also need to be confident that the prevailing means of surveillance will provide them with evidence of noncompliance should their erstwhile partner cheat.

The Washington Naval Conference and the SALT I accords are examples of successful arms control agreements in which leaders were able to make credible commitments to limit central strategic weapons programs precisely because those assets had, by the time of the negotiations, become "known commodities." Technological surprise had effectively been removed from the concerns of statesmen. This does not diminish the importance of the agreements. The Five-Power Treaty significantly improved the security of the three Pacific powers, and also allowed for the reconciliation of political disagreements for almost two decades. The Interim Agreement on Offensive Weapons similarly created a balance between the superpowers that dampened a potential arms race and improved the security of both superpowers. The ABM Treaty was somewhat of an anomaly for MET in that effective strategic defenses would have represented a truly decisive military application. However, given that both superpowers had worked on the problem of ABM defenses for some time and perceived that such a program would be unattainable for some time, the United States and the Soviet Union were able to forge a comprehensive agreement to limit them. In the absence of such an agreement, both states would likely have been forced to continue to devote resources to the development and deployment of these programs, with destabilizing implication for the strategic balance and the underlying relationship of assured destruction that had emerged. Thus, while the technology to create an effective ABM system continues to prove elusive, the competitive dynamic that had emerged in the superpower arms race, and the fear of technological surprise, would have precluded either state from unilaterally deciding to halt ABM development. Thus, the ABM Treaty removed a potentially destabilizing force from the superpower relationship and improved the strategic security environment.

MET is not without limitations, and while it can provide a strong explanation of the failure of the Preparatory Commission and the subsequent World Disarmament Conference, the existence of deep underlying political disputes cannot be ignored. Security-seeking states must be capable of moving beyond bitter political conflicts to achieve any lasting success. Successive French leaders never seemed to view Germany as anything other than a greedy, revisionist state during various negotiations, which made a settlement effectively impossible. Moreover, the hard bargaining exhibited in the "successful" cases of Washington and SALT clearly reflect the inherent difficulty in security cooperation, and also disprove the more pessimistic analyses that equate such success with benign or noncompetitive political international environments.

This book has shown that MET provides a useful approach for understanding international security affairs in general, and for predicting the possibility of success in the persistently difficult area of security cooperation.[1] Specifically, qualitative arms control or arms limitation, as a crucial subset of security cooperation, has received fairly limited interest in the field of international relations. Whereas offensive realism dismisses its importance, and provides little or no explanation for its (even limited) successes, and defensive realism acknowledges the potential importance of arms control, but it would perhaps predict more cooperation than the historical record reflects, MET explains and predicts both the competitive and cooperative impulses that will shape a state's views on cooperation. By incorporating a more robust and comprehensive conceptualization of military technology and technological change improves upon existing realist analyses by explaining the conditions under which arms control is most likely to succeed and where it will fail. Assuming the dynamic nature of technology and focusing on the expectations of leaders provides a more appropriate means of understanding how the prospects for arms control can improve over time or, conversely, how arms control may fail even if existing conditions may be coded as favorable for agreement.

While MET provides a useful theoretical approach for further study of security cooperation, particularly in the realm of arms control, I also believe that it may provide a fruitful means for understanding and explaining the larger issue of military innovation and preparation for war across time and space. These are clearly critical questions for security studies and for international relations in general. Moreover, understanding the interaction between the innovation and development of military technology by states in an anarchic system also has real practical applications for policymakers and security practitioners.

IMPLICATIONS OF MILITARY EXPECTATIONS THEORY

What implications does MET have for current international security challenges? This is a complex and interesting question. Only a decade ago, the United States stood as the single greatest power in the world, based not only upon its share of relative power measured in traditional terms, but also its perceived technological leadership and capacity for innovation. In an increasingly globalized world, technology, or at least the "software" and knowledge components of technology, seems to be diffused at ever more rapid rates. The information technology revolution has had remarkable and wide-ranging implications in the military, civilian, and commercial realms. The United States has clearly benefitted from these developments, particularly in the wake of the collapse of the Soviet Union and the onset of globalization in the 1990s. At the same time, while these trends generally may provide great opportunities for many lesser-developed states in terms of development, they also may create serious challenges in the security realm. What would an MET-based analysis advise to U.S. policymakers in such an environment?

Assessing the Current Security Environment

More recently, the world has become a much more uncertain place. China has continued to modernize its military and is increasingly pressing its interests in disputes with regional neighbors like Japan, the Philippines, and Vietnam.[2] Vladimir Putin has seemingly committed Russia to confrontational policy toward the United States, undermining U.S. interests in Eastern Europe and the Middle East, and expanding Russia's military power.[3] The Middle East is increasingly unstable and the threat of global terrorism remains a central focus of U.S. leaders.[4]

While political uncertainty has increased around the globe, the nature of technology and technological change has also been transforming at a rapid pace. The information technology revolution has dramatically changed the ordinary lives of people around the globe over the past two decades. In the realm of military affairs, we have seen how the United States has been able to harness high technology to increase its ability to project power and the capabilities of precision-guided munitions, directed by networked communications and computers.[5] Even the acronym C2 for "Command and Control" has been expanded to address the various aspects of high technology at work. C4ISR (Command, Control, Communication, Computers, Intelligence, Surveillance, and Reconnaissance) is the key in today's military environment.[6] The seemingly ubiquitous role of unmanned aerial vehicles (UAVs) or "drones" in both American military operations around the world and increasingly for various civilian and commercial applications at home underscores U.S. leadership in cutting-edge technological applications.[7]

However, with its reliance on high technology and advanced networks, new vulnerabilities have emerged that planners fear could be exploited in a future conflict. Given the central role that personal computers, tablets, smartphones, and internet access play in everyday lives, fears that these very assets could be compromised or exploited are increasingly pressing issues for security experts. Moreover, the diffusion of relatively advanced technologies may provide states and nonstate actors with decisive new military capabilities that can seriously threaten both traditional military and civilian targets.[8]

Maintain U.S. Leadership

At first glance, the United States would seem to be in a highly favorable position to maintain its dominant position in global affairs, precisely because of its technological advantages, educated workforce, technical-industrial infrastructure, military prowess, and relative wealth.[9] However, as American resources have declined, the world has become increasingly multipolar, even if states have not utilized more traditional balancing approaches to address U.S. power.[10] In order to maintain its security against the rise of peer competitors, the United States should seek to mitigate costly shorter-term commitments, to focus its energies and resources on maintaining and perpetuating its qualitative advantages. A recalibration of priorities in this direction provides the United States with the greatest probability of successfully exploiting its remaining advantages in relative power to remain at the leading edge of military applications of technological change.

While it is now commonplace to question whether the United States has somehow lost its preeminent position in world politics and whether the unipolar moment is passing, the prescriptive insight of MET for current U.S. political leaders and policy makers is more basic.[11] Maintaining a technological edge requires commitments to investment in basic research in science and technology. The underlying elements of national power that have contributed to the United States as achieving a "Command of the Commons," would place the United States in the best possible position to address new threats and maintain its own security.[12] Many of specific policy prescriptions that flow from this notion are perhaps more domestic in nature, but despite the many challenges that the United States faces in an uncertain world, the greatest challenge may be domestic political dysfunction that threatens to undermine the nation's ability to maintain and expand its technological edge over the long-term.[13] As the case studies of this volume imply, it is often advantageous to be a leader, rather than a laggard in terms of technological breakthroughs. As a great power that is focused on maintaining its security and working with like-minded friends and allies to foster a secure, stable, and prosperous international security environment, the maintenance of a robust technological base should be a vital national interest.

A New Great Power Rivalry?

It is perhaps unsurprising one of the central (and most publicized) characteristics of the growing rivalry between the United States and the People's Republic of China (PRC) is a technological competition.[14] China's decade-long military modernization program has seemingly transformed the People's Liberation Army (PLA) from a mass army to a modern, high-tech fighting force. In some areas, experts even fear that Beijing has surpassed the United States. While this may be due in part to the natural diffusion of technology, claims of large-scale espionage and exploitation of vulnerable American computer networks have exacerbated tensions. The larger, long-term fear is that China could surpass the United States in terms of innovation, technological advancement, and productivity as its indigenous industrial and technical capacities improve.[15]

Much of China's investment in new military applications has focused on obtaining capabilities to address specific needs. On particular focus of investment has been in the realm of short- and medium-range missiles that are meant to clearly deter a unilateral Taiwanese declaration of independence, and to deter others (most notably the United States) from intervening on Taiwan's behalf in the event that a cross-straits conflict broke out.[16] China has also reportedly invested in "information war" capabilities that could limit the ability of the United States to resupply, reinforce, or operate with Taiwan and perhaps Japan in the event of a conflict.[17]

Just as interestingly though, is what China has *not* invested in. It has not expanded its limited ICBM force.[18] Similarly, contrary to fears in the early 1990s, it has not made a major expansion in a blue water navy. These two examples of seeming restraint on the part of Chinese military leaders are clearly predicated on avoiding conflict with the United States in the near-to-medium term, unless in the context of a Taiwan scenario. Chinese military assets thus generally deployed for deterrence and, if necessary, defensive operations.[19] This "asymmetric" approach is generally seen as highly effective, in that it minimizes potential for fears or suspicions in the United States but provides robust assets to counter American forces in the event of a conflict. Where China is clearly a rising power and committed to internal development, this strategic approach makes a great deal of sense, but it is also made possible by China's access to high technology and a commitment to innovation among the Communist Party leadership.

Considering Downsides of Innovation

The pressures created by technological change are strong, and even the remaining superpower is not immune to its influence. The security challenges presented by a rising China, ambiguous Russia, rogue states like Iran and North Korea, and terrorist network with global reach are daunting, and while this period of relative international calm would seem to present an opportunity to initiate and engage in deeper cooperation, political and technological change is inevitable. MET would predict that the United States would continue to leverage upon its existing advantages to develop and deploy new decisive applications to maintain its position of world leadership and improve its security through unilateral means. Embarking on expansive cyber-warfare capabilities, national missile defenses, and the preemptive American militarization of space certainly seem to be early indicators that MET will remain a relevant theoretical approach for quite some time. Space weapons have generally lagged behind the communication, surveillance, and intelligence roles for satellite technology. However, some sort of "killer satellite" would most likely be viewed as a "killer application" the presence of which would dramatically impact the security environment and potentially the relations between the United States and China. More hawkish elements on both sides of the Pacific have advocated for the proactive militarization of space, to "get there first."[20] As MET would predict, the first deployment of a viable anti-satellite weapon is likely to spur an intense arms race, and contribute to the erosion of political and diplomatic relations we have seen in past cases. Similarly, U.S. research and experimentation on hypersonic weapons have seemingly spurred interest in other states like China and Russia, potentially initiating a new arms race in that field of breakthrough application.[21] In short, it may be prudent to consider the likely national security implications of a military technological breakthrough well before its development.

Security Cooperation Can Be Beneficial

Some may argue that it is past time for discussions between the United States and China (as well as other interested states) on the control and management of outer space.[22] While the United States has successfully utilized space for economic, commercial, and military purposes for some time, China has rapidly become a major actor in outer space as well and only looks to be increasing its presence there. In expanding into space, both states have created major vulnerabilities that could be exploited by the other with grave ramifications. Nonetheless, precisely because of these perceived vulnerabilities and the high costs of some sort of extraterrestrial conflict between the United States and the PRC, it may be an appropriate time to attempt to break an emerging security dilemma dynamic through cooperation, namely arms control.[23] This may prove difficult. Certain anti-satellite (ASAT) weapons would seemingly lend themselves to limitation and surveillance, but newer technologies may prove impossible to confidently account for under any regime. Given the obvious benefits of a demilitarized outer space, and the perceived costs of a competition and perhaps military conflict over space, it would seem that leaders in Washington and Beijing should seriously consider cooperative solutions to this challenge.[24]

The larger question for security cooperation remains whether it would make strategic sense and serve U.S. interests for Washington to attempt to bind other emerging powers in arms control agreements that could improve longer-term international security by decreasing or removing incentives for states to seek to acquire weapons, strategic or otherwise. Part of the difficulty of approaching China or other states on outer space is the recent U.S. record of rejecting the Comprehensive Test Ban Treaty (CTBT) and withdrawing from the ABM Treaty.[25] By any objective measure, the United States has signaled a clear preference for unilateralism in the realm of strategic weapons and national security more broadly.

Nonproliferation has been a priority of the United States, and it has worked with allies at both the global and regional levels to remove the perceived need for nuclear weapons and missile technologies to make the costs of noncompliance truly prohibitive.[26] Successful cases of nuclear nonproliferation such as South Africa, Argentina, and Brazil were, at least partially, shaped by the absence of true regional security threats and the perception that the costs of acquiring weapons (in terms of great power sanctions) would be prohibitive.[27] Such sanctions enter the calculations if states do not face a pressing regional threat, as MET would predict. The presence of a regional adversary that could potentially achieve a decisive weapon like a nuclear bomb would make almost all other concerns beyond acquiring a similar capability immaterial. This is sad fact of life in the Middle East, the Subcontinent, and East Asia.[28] In the case of Iran and North Korea, little short of a U.S. security guarantee is likely to end their weapons programs in any conceivable time frame. Similarly, we would expect the strategic rivalry on the subcontinent to continue, perhaps until both states achieve assured destruction capabilities and thus can bargain without fear of a disarming first strike.[29] In each of these cases, until regional security concerns can be resolved, it is highly improbably that Iran would countenance forgoing the bomb.[30] Even then, it is unlikely that a perhaps more secure Iran would ever dismantle its program without direct negotiations and guarantees from the United States.

Security Cooperation Can Be Difficult to Achieve

Nevertheless, as the argument of this book has underscored, qualitative arms control agreements can be difficult to achieve, precisely because of the expectations that military and political leaders develop regarding the impact of new weapons on their security and the feasibility of effectively monitoring compliance in the event of a successful agreement. In some ways, as the so-called New START Treaty reflects, strategic limitations or reduction may be relatively straightforward.[31] Given that nuclear weapons technology has matured and is well-known and understood among the leading powers, this type of agreement fits within the predictions of MET. However, there are likely to be limits to that cooperation, and more optimistic calls for nuclear abolition are unlikely to prove successful.[32] Continuing U.S. investment in potentially large-scale missile defenses will likely place a floor below which Russia is unlikely to cooperatively reduce its strategic deterrent and given China's limited arsenal it is unlikely to take part in serious strategic discussions.[33] Similarly, U.S. research and development of non-nuclear hypersonic weapons, which could constitute a new "Prompt Global Strike" and provide the United States with long-range conventional counterforce weapon, may dampen the enthusiasm of other states to decrease the strategic deterrent while also spurring interest in the potentially decisive capability.[34]

The emerging prospect of "cyber-warfare" seems to be a particularly unlikely realm for effective security cooperation.[35] Whereas arms control and disarmament typically focuses on limits and reductions of actual military applications, current attempts at "Cyber Arms Control" would effectively seek to restrain technology, a virtual impossibility. The nature of critical infrastructure and networked systems makes the actual limitation of physical artifacts effectively impossible. Because the cyber challenges take place in the virtual world, only activities, or processes, or practices can be restrained. This may change over time, but at this point in time, it is difficult to envision the tools that may be available to counter cyber-attacks, as well as the means of attacking.

A historical analogy to current attempts to constrain cyber-warfare may be the 1899 Hague Conference where forward-looking leaders, concerned about changes in military technology sought initially to restrain those applications that states were acquiring to decrease the potential for war.[36] Ultimately, they could not envision the military applications that were going to actually emerge, so they attempted to create prohibitions against certain kinds of conduct or practices, leading to laws of war that failed to restrain belligerents in the subsequent Great War. There may continue to be debates about how truly decisive cyber-warfare could be in the event of a major military conflict between great powers, but what seems clear is that the absence of tangible assets or weapons, makes discussing the concept of control exceedingly difficult. Perhaps as importantly, the prevalence of nonstate actors in the cyber-world makes attribution and thus monitoring and surveillance exceedingly difficult.

Similarly, attempts to restrain the military application of "nanotechnology" are unlikely to prove fruitful.[37] While we may only be on the cusp of what this potentially breakthrough technology could contribute to achieving both security and larger societal objectives, as MET would predict, the combination of a potentially decisive military application coupled with the inherent difficulties of monitoring their development and utilization would make an attempted arms control impossible.

NOTES

1. Other cases that may provide interesting tests of MET include the 1899 Hague Peace Conference, the 1927 Coolidge Naval Conference, the 1930 London Naval Conference, the SALT II agreement, and the 1987 Intermediate Nuclear Forces (INF) Treaty.
2. See, for example, Edward Wong, "On the Chinese Military's New Map, What Borders?" *New York Times*, July 31, 2014.
3. James Miller, "Putin Practiced His Ukraine Strategy in Syria," *Moscow Times*, July 16, 2014.
4. Ammar Al Shamary and Glgamesh Nabeel, "Islamic Extremists Impose Reign of Terror in Iraq," *USA Today*, August 4, 2014.
5. Eliot A. Cohen, "A Revolution in Warfare," *Foreign Affairs* 75, no. 2 (March/April 1996): 37–54.
6. Paul K. Davis, James H. Bigelow, Jimmie McEver, *Effects of Terrain, Maneuver Tactics, and C4ISR on the Effectiveness of Long-Range Precision Fires* (Santa Monica: RAND Corporation, 2000).
7. See, for example, Lynn E. Davis et al, *Armed and Dangerous? UAVs and U.S. Security* (Santa Monica: RAND Corporation, 2014).
8. Andrew F. Krepinevich Jr., "The Pentagon's Wasting Assets: The Eroding Foundations of American Power," *Foreign Affairs* 88, no. 4 (July/August 2009): 18–33.
9. Barry Posen, "Command of the Commons: The Military Foundation of U.S. Hegemony," *International Security* 28, no. 1 (2003).
10. See, for example, Richard N. Haass, "The Age of Nonpolarity: What Will Follow U.S. Dominance," *Foreign Affairs* 87, no. 3 (May/June 2008): 44–56; Joseph S. Nye Jr., "The

Future of American Power: Dominance and Decline in Perspective," *Foreign Affairs* 89, no. 6 (November/December 2010): 2–12; Josef Joffe, "The Default Power: The False Prophecy of America's Decline," *Foreign Affairs* 88, no. 5 (September/October 2009): 21–35; Stephen M. Walt, *Taming American Power: The Global Response to U.S. Primacy* (New York: W. W. Norton, 2005).

11. Charles Krauthammer, "The Unipolar Moment Revisited," *National Interest* 70 (Winter 2002/2003): 5–17.

12. Posen, "Command of the Commons."

13. Fareed Zakaria, "The Future of American Power: How America Can Survive the Rise of the Rest," *Foreign Affairs* 87, no. 3 (May/June 2008): 18–43.

14. Stuart E. Johnson and Duncan Long, eds., *Coping with the Dragon: Essays of PLA Transformation and the U.S. Military* (Washington, DC: National Defense University, 2007).

15. Arvind Subramanian, "The Inevitable Superpower: Why China's Dominance Is a Sure Thing," *Foreign Affairs* 90, no. 5 (September/October 2011): 66–78.

16. John Stillion, "Fighting under Missile Attack," *Air Force Magazine* 92, no. 8 (August 2009): 34–37.

17. Thomas J. Christensen, "China, the U.S.-Japan Alliance, and the Security Dilemma in East Asia," *International Security* 23, no. 4 (1999).

18. Brad Roberts, Robert A. Manning, and Ronald N. Montaperto, "China: The Forgotten Nuclear Power," *Foreign Affairs* 79, no. 4 (2000).

19. Thomas J. Christensen, "Posing Problems without Catching Up: China's Rise and Challenges for U.S. Security Policy," *International Security* 25, no. 4 (2001).

20. Henry F. Cooper and Robert L. Pfaltzgraff Jr., "Lost in Space," *Wall Street Journal*, August 30, 2006.

21. "Hypersonic Missiles: Speed Is the New Stealth," *Economist*, June 1, 2013.

22. Bruce M. DeBlois et al., "Space Weapons: Crossing the U.S. Rubicon," *International Security* 29, no. 2 (2004); William Martel and Toshi Yoshihara, "Averting a Sino-U.S. Space Race," *Washington Quarterly* 26, no. 4 (2003).

23. James Clay Moltz, *The Politics of Space Security: Strategic Restraint and the Pursuit of National Interest* (Stanford, CA: Stanford University Press, 2010).

24. Richard L. Garwin, "Space Weapons or Space Arms Control?," *Proceedings of the American Philosophical Society* 145, no. 3 (2001).

25. On the CTBT, see, for example, Stephen Cambone, "An Inherent Lesson in Arms Control," *The Washington Quarterly* 23, no. 2 (2000); Terry L. Deibel, "The Death of a Treaty," *Foreign Affairs* 81, no. 5 (2002). On the U.S. and the missile defense, see James M. Lindsay and Michael E. O'Hanlon, "Missile Defense after the ABM Treaty," *Washington Quarterly* 25, no. 3 (2002); James M. Lindsay et al., "Limited National and Allied Missile Defense," *International Security* 26, no. 4 (2002); John Newhouse, "The Missile Defense Debate," *Foreign Affairs* 80, no. 4 (2001).

26. Peter D. Feaver, Scott D. Sagan, and David J. Karl, "Proliferation Pessimism and Emerging Nuclear Powers," *International Security* 22, no. 2 (1997); Scott D. Sagan, "Why Do States Build Nuclear Weapons?: Three Models in Search of a Bomb," *International Security* 21, no. 3 (1996).

27. On nonproliferation, see Ariel E. Levite, "Never Say Never Again: Nuclear Reversal Revisited," *International Security* 27, no. 3 (2002); Peter Liberman, "The Rise and Fall of the South African Bomb," *International Security* 26, no. 2 (2001); T. V. Paul, *Power Versus Prudence: Why Nations Forgo Nuclear Weapons* (Montreal: McGill-Queens University Press, 2000).

28. On India and Pakistan, see K. Shankar Bajpai, "Untangling India and Pakistan," *Foreign Affairs* 82, no. 3 (2003); Strobe Talbott, "Dealing with the Bomb in South Asia," *Foreign Affairs* 78, no. 2 (1999).

29. Paul S. Kapur, "India and Pakistan's Unstable Peace: Why Nuclear South Asia Is Not Like Cold War Europe," *International Security* 30, no. 2 (Fall 2005): 127–52.

30. Shahram Chubin and Robert S. Litwak, "Debating Iran's Nuclear Ambitions," *Washington Quarterly* 26, no. 4 (2003); Kenneth Pollack and Ray Takeyh, "Taking on Tehran," *Foreign*

Affairs 84, no. 2 (2005); Scott D. Sagan, "How to Keep the Bomb from Iran," *Foreign Affairs* 85, no. 5 (2006).

31. On New START, see Jonathan Weisman, "U.S. and Russia Seek More Extensive Weapons Cuts," *Wall Street Journal*, July 2, 2009. "Joint Understanding for the Start Follow on Treaty," (The White House, Office of the Press Secretary, July 8, 2009).

32. George Perkovich and James Acton, *Abolishing Nuclear Weapons* (London: IISS, 2008).

33. On Russia, see Alexei Arbatov, *Gambit or Endgame? The New State of Arms Control* (Washington, DC: Carnegie Endowment for International Peace, 2011). On China, see: M. Taylor Fravel and Evan S. Medeiros, "China's Search for Assured Retaliation: The Evolution of Chinese Nuclear Strategy and Force Structure," *International Security* 35, no. 2 (2010), 48–87; Lora Saalman, *China and the U.S. Nuclear Posture Review* (Washington, DC: Carnegie Endowment for International Peace, 2011).

34. On "Prompt Global Strike," see Bruce M. Sugden, "Speed Kills: Analyzing the Development of Conventional Ballistic Missiles," *International Security* 34, no. 1 (Summer 2009); 113–46; Austin Long, Dinshaw Mistry, Bruce M. Sugden, "Going Nowhere Fast: Assessing Concerns about Long-Range Conventional Missiles," *International Security* 34, no. 4 (Spring 2010): 166–84.

35. See, for example, Lucas Kello, "The Meaning of the Cyber Revolution: Perils to Theory and Statecraft," *International Security* 38, no. 2 (Fall 2013): 7–40; Martin R. Stytz and Sheila B. Banks, "Toward Attaining Cyber Dominance," *Strategic Studies Quarterly* 8, no. 1 (Spring 2013): 55–87. For more skeptical views, see: Erik Gartzke, "The Myth of Cyberwar: Bringing War in Cyberspace Back Down to Earth," *International Security* 38, no. 2 (Fall 2013): 41–73. Thomas Rid, "Cyber War Will Not Take Place," *Journal of Strategic Studies* 35, no. 1 (February 2012): 5–32. Martin C. Libicki, "Why Cyber War Will Not and Should Not Have Its Grand Strategist," *Strategic Studies Quarterly* 8, no. 1 (Spring 2014): 23–39.

36. Calvin DeArmond Davis, *The United States and the First Hague Peace Conference* (Ithaca, NY: Cornell University Press, 1962).

37. Jurgen Altmann, "Military Uses of Nanotechnology: Perspectives and Concerns," *Security Dialogue* 35, no. 1 (2004): 61–79.

Bibliography

Acheson, Dean. *Present at the Creation: My Years in the State Department*. New York: W. W. Norton, 1987.

Adams, Karen Ruth. "Attack and Conquer? International Anarchy and the Offense-Defense-Deterrence Balance." *International Security* 28, no. 3 (2003/2004): 45–83.

Allen, Scott. "A Comparison of the Washington Naval Arms Limitation Treaty of 1922 and the Strategic Arms Limitation Agreements of 1972." PhD Dissertation, University of Hawaii, 1976.

Allison, Graham T., and Frederic A. Morris. "Armaments and Arms Control: Exploring the Determinants of Military Weapons." In *Arms, Defense Policy, and Arms Control*, edited by Franklin A. Long and George W. Rathjens, 99–130. New York: W. W. Norton, 1976.

Alperovitz, Gar. *Atomic Diplomacy: Hiroshima and Potsdam, the Use of the Atomic Bomb and the American Confrontation with Soviet Power*. New York: Vintage Books, 1965.

Altmann, Jurgen. "Military Uses of Nanotechnology: Perspectives and Concerns." *Security Dialogue* 35, no. 1 (2004): 61–79.

Arbatov, Georgi. *The System: An Insider's Life in Soviet Politics*. New York: Times Books, 1992.

"Arms Control to Improve American Security, Address by Gerard Smith." *Department of State Bulletin* 67, no. 1736 (1972).

Armstrong, David. "China's Place in the New Pacific Order." In *The Washington Conference, 1921–22: Naval Rivalry, East Asian Stability and the Road to Pearl Harbor*, edited by Erik Goldstein and John Maurer, 249–66. Essex: Frank Cass, 1994.

Art, Robert J. "To What Ends Military Power?" *International Security* 4, no. 4 (1980): 3–35.

Asada, Sadao. *From Mahan to Pearl Harbor: The Imperial Japanese Navy and the United States*. Annapolis, MD: Naval Institute Press, 2006.

———. "From Washington to London: The Imperial Japanese Navy and the Politics of Naval Limitation, 1921–30." In *The Washington Conference, 1921–22: Naval Rivalry, East Asia Stability and the Road to Pearl Harbor*, edited by Erik Goldstein and John Maurer, 147–91. Portland: Frank Cass, 1994.

———. "Japan's 'Special Interests' and the Washington Conference." *The American Historical Review* 61, no. 1 (1961): 62–70.

Axelrod, Robert. *The Evolution of Cooperation*. New York: Basic Books, 1984.

Axelrod, Robert, and Robert O. Keohane. "Achieving Cooperation under Anarchy: Strategies and Institutions." *World Politics* 38, no. 1 (1985): 226–54.

Bade, Robin, and Michael Parkin, eds. *Foundations of Economic*. 2nd ed. Boston: Pearson Addison-Wesley, 2004.

Bailey, Jonathan B. A. "The First World War and the birth of modern warfare." In *The Dynamics of Military Revolution*, edited by MacGregor Knox and Williamson Murray, 132–53. Cambridge: Cambridge University Press, 2001.

Bajpai, K. Shankar. "Untangling India and Pakistan." *Foreign Affairs* 82, no. 3 (2003).

Ball, Desmond. *Politics and Force Levels: The Strategic Missile Program of the Kennedy Administration*. Berkeley: University of California Press, 1980.

Barton, John H., and Lawrence D. Weiler, eds. *International Arms Control: Issues and Agreements*. Stanford, CA: Stanford University Press, 1976.

Baruch, Bernard M. *Baruch: The Public Years*. New York: Holt, Rinehart, and Winston, 1960.

Basalla, George. *The Evolution of Technology*. Cambridge: Cambridge University Press, 1989.

Bayliss, John, and Gerald Segal, eds. *Soviet Strategy*. London: Croom Helm, 1981.

Bechhoefer, Bernard C. *Postwar Negotiations for Arms Control*. New York: Greenwood Press, 1961.

Beisner, Robert L. *Dean Acheson: A Life in the Cold War*. Oxford: Oxford University Press, 2006.

Bell, Coral. *The Diplomacy of Detente: The Kissinger Era*. New York: St. Martin's Press, 1977.

Bennett, Geoffrey. *The Battle of Jutland*. Hertfordshire: Wordsworth Editions, 1999.

Berman, Robert P., and John C. Baker. *Soviet Strategic Forces: Requirements and Responses*. Washington, DC: Brookings Institution, 1982.

Bernstein, Robert A., and William W. Anthony. "The ABM Issue in the Senate, 1968–1970: The Importance of Ideology." *American Political Science Review* 68, no. 3 (1974): 1198–206.

Betts, Richard K. "Must War Find a Way?" *International Security* 24, no. 2 (1999): 166–98.

———. *Nuclear Blackmail and Nuclear Balance*. Washington, DC: Brooking Institution Press, 1987.

———. "A Nuclear Golden Age? The Balance before Parity." *International Security* 11, no. 3 (1986/1987): 3–32.

Beyerchen, Alan. "Clausewitz, Nonlinearity, and the Unpredictability of War." *International Security* 17, no. 3 (1992–1993): 59–90.

Biddle, Stephen. *Military Power: Explaining Victory and Defeat in Modern Battle*. Princeton, NJ: Princeton University Press, 2004.

———. "Rebuilding the Foundations of Offense-Defense Theory." *Journal of Politics* 63, no. 3 (2001): 741–74.

Bijker, Wiebe E., Thomas P. Hughes, and Trevor Pinch, eds. *The Social Construction of Technological Change: New Directions in the Sociology and History of Technology*. Cambridge, MA: MIT Press, 1989.

Bijker, Wiebe E., and John Law, eds. *Shaping Technology/Building Society: Studies in Societal Change*. Cambridge, MA: MIT Press, 1994.

Blacker, Coit D., and Gloria Duffy, eds. *International Arms Control: Issues and Agreements*. 2nd ed. Stanford, CA: Stanford University Press, 1984.

Blackett, P. M. S. *Fear, War, and the Bomb: Military and Political Consequences of Atomic Energy*. New York: Whittlesey House, 1949.

Blatt, Joel. "France and the Washington Conference." In *The Washington Conference, 1921–22: Naval Rivalry, East Asian Stability and the Road to Pearl Harbor*, edited by Erik Goldstein and John Maurer, 192–219. Essex: Frank Cass, 1994.

Bluth, Christoph. *Soviet Strategic Arms Policy before Salt*. Cambridge: Cambridge University Press, 1992.

Boggs, Marion Willam. *Attempts to Define and Limit "Aggressive" Armaments in Diplomacy and Strategy*. Columbia: University of Missouri Press, 1941.

Bond, Brian, and Martin Alexander, "Liddell Hart and De Gaulle: The Doctrines of Limited Liability and Mobile Defense," in *Makers of Modern Strategy from Machiavelli to the Nuclear*, edited by Peter Paret, 598–623. Princeton, NJ: Princeton University Press, 1986.

Bonney, George. *The Battle of Jutland 1916*. London: Sutton Publishing, 2006.

Bottome, Edgar M. *The Missile Gap: A Study of the Formulation of Military and Political Policy*. Cranbury, NJ: Fairleigh Dickinson University Press, 1971.

Bowker, Mike. "Brezhnev and Superpower Relations." In *Brezhnev Reconsidered*, edited by Edwin Bacon and Mark Sandle, 90–109. Hampshire: Palgrave Macmillan, 2002.

Braisted, William Reynolds. *The United States Navy in the Pacific, 1900–1922*. Austin: University of Texas Press 1971.

Bresler, Robert J., and Robert C. Gray. "The Bargaining Chip and Salt." *Political Science Quarterly* 92, no. 1 (1977): 65–88.

Brodie, Bernard, and Fawn Brodie. *From Crossbow to H-Bomb*. New York: Dell Publishing, 1962.

Buckley, Thomas H. "The Icarus Factor: The American Pursuit of Myth in Naval Arms Control, 1921–1930." In *The Washington Conference, 1921–22: Naval Rivalry, East Asia Stability and the Road to Pearl Harbor*, edited by Erik Goldstein and John Maurer. Portland: Frank Cass, 1994.

———. *The United States and the Washington Conference, 1921–1922*. Knoxville: University of Tennessee Press, 1970.

Buell, Raymond Leslie. *The Washington Conference*. New York: D. Appleton, 1922.

Buhite, Russell D., and W. Christpoher Hamel. "War for Peace: The Question of an American Preventive War against the Soviet Union, 1945–1955." *Diplomatic History* 14, no. 3 (1990): 367–84.

Bull, Hedley. *The Control of the Arms Race: Disarmament and Arms Control in the Missile Age*. New York: Praeger Publishers, for the International Institute for Strategic Studies, 1961.

Bullock, Alan. *Hitler and Stalin: Parallel Lives*. New York: Alfred A. Knopf, 1992.

Bundy, McGeorge. *Danger and Survival: Choices about the Bomb in the First Fifty Years*. New York: Random House, 1988.

Bunn, George. *Arms Control by Committee: Managing Negotiations with the Russians*. Stanford, CA: Stanford University Press, 1992.

Butler, Daniel Allen. *Distant Victory: The Battle of Jutland and the Allied Triumph in the First World War*. London: Praeger, 2006.

Buzan, Barry, and Eric Herring. *The Arms Dynamic in World Politics*. Boulder: Lynne Rienner, 1998.

Byrnes, James F. *Speaking Frankly*. New York: Harper and Brothers, 1947.

Cambone, Stephen. "An Inherent Lesson in Arms Control." *The Washington Quarterly* 23, no. 2 (2000): 207–18.

Campbell, John. *Jutland: An Analysis of the Fighting*. London: Lyons Press, 1998.

Carnesale, Albert, and Richard N. Haass, eds. *Superpower Arms Control: Setting the Record Straight*. Cambridge: Ballinger Publishing Company, 1987.

Carter, Ashton B., and David N. Schwartz, eds. *Ballistic Missile Defense*. Washington, DC: Brookings Institution, 1984.

Caspary, William R. "Richardson's Model of Arms Races: Description, Critique, and an Alternative Model." *International Studies Quarterly* 11, no. 1 (1967): 63–88.

Catudal, Honore M. *Soviet Nuclear Strategy from Stalin to Gorbachev: A Revolution in Soviet Military and Political Thinking*. Atlantic Highlands: Humanities Press International, 1988.

Cha, Victor D. "Korea's Place in the Axis." *Foreign Affairs* 81, no. 3 (2002).

Christensen, Thomas J. "China, the U.S.-Japan Alliance, and the Security Dilemma in East Asia." *International Security* 23, no. 4 (1999): 49–80.

———. "Posing Problems without Catching Up: China's Rise and Challenges for U.S. Security Policy." *International Security* 25, no. 4 (2001): 5–40.

Chubin, Shahram, and Robert S. Litwak. "Debating Iran's Nuclear Ambitions." *Washington Quarterly* 26, no. 4 (2003).

Citino, Robert M. *Path to Blitzkrieg: Doctrine and Training in the German Army, 1920–39*. Mechanicsburg, PA: Stackpole Books, 2008.

———. *Quest for Decisive Victory: From Stalemate to Blitzkrieg in Europe, 1899–1940*. Lawrence: University of Kansas Press, 2002.

Clearwater, John M. *Johnson, Mcnamara, and the Birth of Salt and the ABM Treaty 1963–1969*: Dissertation.com, 1996.

Cloud, David S. "Officials Call Test a Success after a Missile Is Shot Down." *New York Times*, September 2, 2006, A11.
Cohen, Eliot A., and John Gooch. *Military Misfortunes: The Anatomy of Failure in War*. New York: Vintage Books, 1990.
Collins, John M. *American and Soviet Military Trends since the Cuban Missile Crisis*. Washington, DC: Center for Strategic and International Studies, 1978.
Committee on Armed Services, U.S. House of Representatives. *Military Posture*, 91st Cong., 2nd Sess., February–April 1970.
Committee on Foreign Relations, U.S. Senate. *Conference on the Limitation of Armament*, 67th Cong. 2nd Sess., 1922.
Committee on Naval Affairs, U.S. House of Representatives. *Naval Policy of the United States*, 66th Congress, 3rd Session, 1921.
Cooper, Henry F., and Robert L. Pfaltzgraff Jr. "Lost in Space." *Wall Street Journal*, August 30, 2006.
Copeland, Dale C. *The Origins of Major War*. Ithaca, NY: Cornell University Press, 2000.
———. "Review: The Constructivist Challenge to Structural Realism: A Review Essay." *Review of Social Theory of International Politics* by Wendt, Alexander. *International Security* 25, no. 2 (2000): 187–212.
———. "Theory and History in the Study of Major War." *Security Studies* 10, no. 4 (2001): 212–39.
Corum, James S. *The Roots of Blitzkrieg: Hans Von Seeckt and German Military Reform*. Lawrence: University Press of Kansas, 1994.
Daalder, Ivo H. "The Limited Test Ban Treaty." In *Superpower Arms Control: Setting the Record Straight*, edited by Albert Carnesale and Richard N. Haass, 9–41. Cambridge: Ballinger, 1987.
Dalleck, Robert. *Nixon and Kissinger: Partners in Power*. New York: HarperCollins, 2007.
Dallin, Alexander, ed. *The Soviet Union, Arms Control, and Disarmament: A Study of Soviet Attitudes*. New York: School of International Affairs, Columbia University, 1964.
Davis, James W., Jr., et al. "Taking Offense at Offense-Defense Theory." *International Security* 23, no. 3 (1998): 179–206.
Davis, Lynn Etheridge, and Warner R. Schilling. "All You Ever Wanted to Know about MIRV and ICBM Calculations but Were Not Cleared to Ask." *Journal of Conflict Resolution* 17, no. 2 (1973): 207–42.
Deardorff, Alan V. "Growth Paths in the Solow Neoclassical Growth Model." *Quarterly Journal of Economics* 84, no. 1 (1970): 134–39.
DeBlois, Bruce M., et al. "Space Weapons: Crossing the U.S. Rubicon." *International Security* 29, no. 2 (2004): 50–84.
Deibel, Terry L. "The Death of a Treaty." *Foreign Affairs* 81, no. 5 (2002).
Diehl, Paul F. "Armaments without War: An Analysis of Some Underlying Effects." *Journal of Peace Research* 22, no. 3 (1985): 249–59.
———. "Arms Races and Escalation: A Closer Look." *Journal of Peace Research* 20, no. 3 (1983): 205–12.
———. "Arms Races to War: Testing Some Empirical Linkages." *The Sociological Quarterly* 26, no. 3 (1985): 331–49.
———. "Ghosts of Arms Control Past." *Political Science Quarterly* 105, no. 4 (1990): 597–615.
Diehl, Paul F., and Mark J. C. Crescenzi. "Reconfiguring the Arms Race-War Debate." *Journal of Peace Research* 35, no. 1 (1998): 111–18.
Dingman, Roger. *Power in the Pacific: The Origin of Naval Arms Limitation, 1914–1922*. Chicago: University of Chicago Press, 1976.
Committee on Naval Affairs, US House of Representatives. *Disarmament*, 66th Cong., 3rd Sess., 1921.
Committee on Foreign Affairs, US House. *Disarmament*, 66th Congress, 3rd Sess., 1921.
Dobrynin, Anatoly. *In Confidence: Moscow's Ambassador to America's Six Cold War Presidents*. Seattle: University of Washington Press, 1995.

Documents on British Foreign Policy, 1919–1939: Volume Xiv, Far Eastern Affairs, April 1920–February 1922. Edited by Great Britain Foreign Office. Oxford: Her Majesty's Stationary Office, 1966.

Documents on Disarmament. Washington, DC: United States Arms Control and Disarmament Agency, 1971.

Documents on Disarmament, 1945–1959. Washington, DC: U.S. Department of State, 1960.

Douglass, Joseph D., and Amoretta M. Hoeber. *Soviet Strategy for Nuclear War*. Stanford, CA: Hoover Institution Press, 1979.

Downs, George W., David M. Rocke, and Randolph M. Siverson. "Arms Races and Cooperation." *World Politics* 38, no. 1 (1985): 118–46.

Enthoven, Alain C., and K. Wayne Smith. *How Much Is Enough? Shaping the Defense Program 1961–1969*. Santa Monica: RAND Corporation, 2005.

Erickson, John, ed. *The Military-Technical Revolution: Its Impact on Strategy and Foreign Policy*. New York: Frederich A. Praeger Publishers, for the Institute for the Study of the USSR, 1966.

Evangelista, Matthew. *Innovation and the Arms Race: How the United States and the Soviet Union Develop New Military Technologies*. Ithaca, NY: Cornell University Press, 1988.

Evans, David C., and Mark R. Peattie. *Kaigun: Strategy, Tactics, and Technology in the Imperial Japanese Navy 1887–1941*. Annapolis, MD: Naval Institute Press, 1997.

Farley, Philip J. "Strategic Arms Control, 1967–1987." In *U.S.-Soviet Security Cooperation: Achievements, Failures, Lessons*, edited by Alexander L. George, Philip J. Farley and Alexander Dallin, 215–53. New York: Oxford University Press, 1988.

Fearon, James D. "Rationalist Explanations for War." *International Organization* 49, no. 3 (1995): 379–414.

Feaver, Peter D. "Blowback: Information Warfare and the Dynamic of Coercion." *Security Studies* 7, no. 4 (1998): 88–120.

Feaver, Peter D., Scott D. Sagan, and David J. Karl. "Proliferation Pessimism and Emerging Nuclear Powers." *International Security* 22, no. 2 (1997): 185–207.

Foreign Relations of the United States, 1922. Vol. 1. Washington, DC: U.S. Department of State.

Foreign Relations of the United States, 1945, Europe. Vol. 5. Washington, DC: U.S. Department of State.

Foreign Relations of the United States, 1945, General Political and Economic Matters. Vol. 2. Washington, DC: U.S. Department of State.

Foreign Relations of the United States, 1946, Eastern Europe, the Soviet Union. Vol. 6. Washington, DC: U.S. Department of State.

Foreign Relations of the United States, 1946, General: The United Nations. Vol. 1. Washington, DC: U.S. Department of State.

Foreign Relations of the United States, 1947, General: The United Nations. Vol. 1. Washington, DC: U.S. Department of State.

Foreign Relations of the United States, 1964–1968, Arms Control and Disarmament. Vol. 11. Washington, DC: U.S. Department of State.

Foreign Relations of the United States, 1964–1968, National Security Policy. Vol. 10. Washington, DC: U.S. Department of State.

Foreign Relations of the United States, 1969–1976, Foundations of Foreign Policy, 1969–1972. Vol. 1. Washington, DC: U.S. Department of State.

Foreign Relations of the United States, 1969–1976, the Soviet Union. Vol. 14. Washington, DC: U.S. Department of State.

Foreign Relations of the United States, 1969–1976, the Soviet Union. Vol. 12. Washington, DC: U.S. Department of State.

Freedman, Lawrence. *The Evolution of Nuclear Strategy*. New York: St. Martin's Press, 1981.

———. *Kennedy's Wars: Berlin, Cuba, Laos, and Vietnam*. Oxford: Oxford University Press, 2000.

French, David. *Raising Churchill's Army: The British Army and the War against Germany*. Oxford: Oxford University Press, 2000.

Frye, Alton. *A Responsible Congress: The Politics of National Security*. New York: McGraw-Hill Book Company for the Council on Foreign Relations, 1975.
Fuller, J. F. C. *Armament and History: The Influence of Armament on History from the Dawn of Classical Warfare to the End of the Second World War*. New York: Da Capo Press, 1998.
Gaddis, John Lewis. *Strategies of Containment: A Critical Appraisal of American National Security Policy during the Cold War*. Revised and Expanded ed. New York: Oxford University Press, 2005.
———. *The United States and the Origins of the Cold War, 1941–1947*. New York: Columbia University Press, 1972.
———. *We Now Know: Rethinking Cold War History*. Oxford: Oxford University Press for the Council on Relations, 1997.
Gamble, Raymond C. "Decline of the Dreadnought: Britain and the Washington Naval Conference, 1921–1922." PhD Dissertation, University of Massachusetts, 1993.
Gardner, Lloyd C. *Architects of Illusion: Men and Ideas in American Foreign Policy, 1941–1949*. Chicago: Quadrangle, 1970.
Garthoff, Raymond L. *Detente and Confrontation: American-Soviet Relations from Nixon to Reagan*. Washington, DC: Brookings Institution, 1985.
———. "Mutual Deterrence and Strategic Arms Limitation in Soviet Policy." *International Security* 3, no. 1 (1978): 112–47.
———. "Mutual Deterrence, Parity and Strategic Arms Limitation in Soviet Policy." In *Soviet Military Thinking*, edited by Derek Leebaert, 92–124. London: George Allen and Unwin, 1981.
———. *Soviet Strategy in the Nuclear Age*. New York: Frederick A. Praeger, 1958.
Garwin, Richard L. "Space Weapons or Space Arms Control?" *Proceedings of the American Philosophical Society* 145, no. 3 (2001): 243–59.
Gelber, Harry G. "Technical Innovation and Arms Control." *World Politics* 26, no. 4 (1974): 509–41.
George, Alexander L., and Andrew Bennett. *Case Studies and Theory Development in the Social Sciences*. Cambridge, MA: MIT Press, 2004.
George, Alexander L., Philip J. Farley, and Alexander Dallin, eds. *U.S.-Security Cooperation: Achievements, Failures, Lessons*. New York: Oxford University Press, 1988.
George, Alexander L., and Timothy J. McKeown. "Case Studies and Theories of Organizational Decision Making." In *Advances in Information Processing in Organizations*, edited by Robert F. Coulam and Richard A. Smith, 21–58. Greenwich: JAI, 1985.
Gerber, Larry G. "The Baruch Plan and the Origins of the Cold War." *Diplomatic History* 6 (1982): 60–95.
Geyer, Michael. "German Strategy in the Age of Machine Warfare, 1914–1945," in *Makers of Modern Strategy from Machiavelli to the Nuclear*, edited by Peter Paret, 527–97. Princeton, NJ: Princeton University Press, 1986.
Giddens, Anthony. *The Nation States and Violence: Volume 2 of a Contemporary Critique of Historical Materialism*. Berkeley: University of California Press, 1987.
Gilpin, Robert. *American Scientists and Nuclear Weapons Policy*. Princeton, NJ: Princeton University Press, 1962.
———. *War and Change in World Politics*. Cambridge: Cambridge University Press, 1981.
Glaser, Charles L. *Rational Theory of International Politics: The Logic of Competition and Cooperation*. Princeton, NJ: Princeton University Press, 2010.
———. "The Causes and Consequences of Arms Racing." *Annual Review of Political Science*, no. 3 (2003): 251–76.
———. "Political Consequences of Military Strategy: Expanding and Refining the Spiral and Deterrence Models." *World Politics* 44, no. 4 (1992): 497–538.
———. "Realists as Optimists: Cooperation as Self-Help." *International Security* 19, no. 3 (1994/95): 50–90.
———. "The Security Dilemma Revisited." *World Politics* 50, no. 1 (1997): 171–201.
———. "When Are Arms Races Dangerous? Rational Versus Suboptimal Arming." *International Security* 28, no. 4 (2004): 44–84.

Glaser, Charles L., and Chaim Kaufmann. "What Is the Offense-Defense Balance and Can We Measure It?" *International Security* 22, no. 4 (1998): 44–82.

Glynn, Patrick. *Closing Pandora's Box: Arms Races, Arms Control and the History of the Cold War*. New York: New Republic Books, 1992.

Goldblatt, Jozef. *Arms Control: The New Guide to Negotiations and Agreements*. London: Sage, 2002.

Goldman, Emily O. *Sunken Treaties: Naval Arms Control between the Wars*. University Park: Pennsylvania State University Press, 1994.

Goldman, Emily O., and Richard B Andres. "Systemic Effects of Military Innovation and Diffusion." *Security Studies* 8, no. 4 (1999): 79–125.

Goldman, Emily O., and Leslie C. Eliason, eds. *The Diffusion of Military Technology and Ideas*. Stanford, CA: Stanford University Press, 2003.

Goldstein, Erik. "The Evolution of British Diplomatic Strategy for the Washington Conference." In *The Washington Conference, 1921–22: Naval Rivalry, East Asia Stability and the Road to Pearl Harbor*, edited by Erik Goldstein and John Maurer, 4–34. Portland: Frank Cass, 1994.

Goldstein, Erik, and John Maurer, eds. *The Washington Conference, 1921–22: Naval Rivalry, East Asia Stability and the Road to Pearl Harbor*. Portland: Frank Cass, 1994.

Graham Fry, Michael. "The Pacific Dominions and the Washington Conference, 1921–22." In *The Washington Conference, 1921–22: Naval Rivalry, East Asian Stability and the Road to Pearl Harbor*, edited by Erik Goldstein and John Maurer, 60–101. Essex: Frank Cass, 1994.

Gray, Colin S. "The Arms Race Phenomenon." *World Politics* 24, no. 1 (1971): 39–79.

———. "The Urge to Compete: Rationales for Arms Racing." *World Politics* 26, no. 2 (1974): 207–33.

Gray, Collin S. *House of Cards: Why Arms Control Must Fail*. Ithaca, NY: Cornell University Press, 1992.

———. *Weapons Don't Make War: Politics, Strategy, and Military Technology*. Lawrence: University Press of Kansas, 1993.

Greenwood, Ted. *Making the MIRV: A Study in Defense Decision Making*. Lanham, MD: University Press of America, 1988.

———. "Why Military Technology Is Difficult to Restrain." *Science, Technology and Human Values* 15, no. 4 (1990): 412–29.

Grieco, Joseph M. "Anarchy and the Limits of Cooperation: A Realist Critique of the Newest Liberal Institutionalism." *International Organization* 42, no. 3 (1988): 485–507.

Grieco, Joseph, Robert Powell, and Duncan Snidal. "The Relative-Gains Problem for International Cooperation." *American Political Science Review* 87, no. 3 (1993): 727–43.

Gromyko, Andrei. *Memoirs*. New York: Doubleday, 1989.

Guderian, Heinz. *Panzer Leader*. New York: De Capo Press, 1996.

Habeck, Mary R. *Storm of Steel: The Development of Armor Doctrine in Germany and the Soviet Union, 1919–1939*. Ithaca, NY: Cornell University Press, 2003.

Hackett, James T. "Missile Defense at 25." *Washington Times*, March 23, 2008, B01.

Haftendorn, Helga, Robert O. Keohane, and Celeste A. Wallander, eds. *Imperfect Unions: Security Institutions over Time and Space*. Oxford: Oxford University Press, 1999.

Hagan, Kennth J. *This People's Navy: The Making of American Seapower*. New York: Touchstone, 1991.

Halperin, Morton H. "The Decision to Deploy the ABM: Bureaucratic and Domestic Politics in the Johnson Administration." *World Politics* 25, no. 1 (1972): 62–95.

Headrick, Daniel R. *The Tools of Empire: Technology and European Imperialism in the Nineteenth Century*. New York: Oxford University Press, 1981.

House of Representatives, Committee on Armed Services. *Hearings on Military Posture*, 91st Cong., 2nd Sess., 1970.

Committee on Foreign Relations, US Senate. *Hearings on Strategic Arms Limitation Agreements*, 92nd Cong., 2nd Sess., June–July 1972.

Herken, Gregg. *The Winning Weapon: The Atomic Bomb in the Cold War 1945–1950*. New York: Random House, 1980.

Herrera, Geoffrey L. "The Mobility of Power: Technology, Diffusion and International Change." PhD Dissertation, Princeton University, 1995.
———. *Technology and International Transformation: The Railroad, the Atom Bomb, and the Politics of Technological Change*. Albany: State University of New York Press, 2006.
Hersh, Seymour M. *The Price of Power: Kissinger in the Nixon White House*. New York: Summit Books, 1983.
Herwig, Holger H. "The Battlefleet Revolution, 1885–1914." In *The Dynamics of Military Revolution 1300–2050*, edited by MacGregor Knox and Williamson Murray, 114–31. Cambridge: Cambridge University Press, 2001.
———."Innovation Ignored: The Submarine Problem, Germany, Britain and the United States, 1919–1939." In *Military Innovation in the Interwar Period*, edited by Williamson Murray and Allan R. Millett, 227–64. Cambridge: Cambridge University Press, 1996.
Hewlett, Richard G., and Oscar E. Anderson, Jr. *The New World, 1939/1946: A History of the United States Atomic Energy Commission Volume I*. University Park: Pennsylvania State University Press, 1962.
Hofman, George F. *Through Mobility We Conquer: The Mechanization of U.S. Cavalry*. Lexington: University Press of Kentucky, 2006.
Holloway, David. *The Soviet Union and the Arms Race*. New Haven, CT: Yale University Press, 1983.
———. "The Soviet Union and the Origins of the Arms Race." In *Origins of the Cold War: An International History*, edited by Melvyn P. Leffler and David S. Painter, 95–106. London: Routledge, 1994.
———. *Stalin and the Bomb: The Soviet Union and Atomic Energy 1939–1956*. New Haven, CT: Yale University Press, 1994.
Hughes, Thomas P. *Human-Built World: How to Think about Technology and Culture*. Chicago: University of Chicago Press, 2005.
Huntington, Samuel P. "Arms Races: Prerequisites and Results." In *The Use of Force*, edited by Robert J. Art and Robert Jervis, 85–118. Lanham, MD: University Press of America, 1993.
———. *The Common Defense: Strategic Programs in National Politics*. New York: Columbia University Press, 1966.
Hurley, Alfred F. *Billy Mitchell: Crusader for Air Power*. Bloomington: Indiana University Press, 2006.
Ichihashi, Yamato. *The Washington Conference and after, a Historical Survey*. Stanford, CA: Stanford University Press, 1928.
International Control of Atomic Energy: Growth of a Policy. Washington, DC: U.S. Department of State, 1946.
Intriligator, Michael D. "Strategic Considerations in the Richardson Model of Arms Races." *Journal of Political Economy* 83, no. 2 (1975): 339–54.
Isaacson, Walter. *Kissinger: A Biography*. New York: Simon and Schuster, 2005.
Jervis, Robert. "Arms Control, Stability, and Causes of War." *Political Science Quarterly* 108, no. 2 (1993): 239–53.
———. "Cooperation under the Security Dilemma." *World Politics* 30, no. 2 (1978): 167–214.
———. "From Balance to Concert: A Study of International Security Cooperation." *World Politics* 38, no. 1 (1985): 58–79.
———. *The Meaning of the Nuclear Revolution: Statecraft and the Prospect of Armageddon*. Ithaca, NY: Cornell University Press, 1989.
———. *Perception and Misperception in World Politics*. Princeton, NJ: Princeton University Press, 1976.
———. "Realism, Game Theory, and Cooperation." *World Politics* 40, no. 3 (1988): 317–49.
———. "Realism, Neoliberalism, and Cooperation: Understanding the Debate." *International Security* 24, no. 1 (1999): 42–63.
———. "Security Regimes." *International Organization* 36, no. 2 (1982): 357–78.
Johnson, David E. *Fast Tanks and Heavy Bombers: Innovation in the U.S. Army, 1917–1945*. Ithaca, NY: Cornell University Press, 1998.

Johnson, Lyndon Baines. *The Vantage Point: Perspectives of the Presidency 1963–1969*. New York: Holt, Rinehart, and Winston, 1971.

Jones, Archer, and Andrew J. Keogh. "The Dreadnought Revolution: Another Look." *Military Affairs* 49, no. 3 (1985): 124–31.

Jungk, Robert. *Brighter Than a Thousand Suns: A Personal History of the Atomic Scientists*. New York: Harcourt Brace, 1958.

Kahan, Jerome H. *Security in the Nuclear Age: Developing U.S. Strategic Arms Policy*. Washington, DC: Brookings Institution, 1975.

Kalb, Marvin, and Bernard Kalb. *Kissinger*. Boston: Little, Brown, 1974.

Kennan, George. *The Decision to Intervene*. New York: Atheneum, 1967.

Kennedy, Paul M. *Grand Strategies in War and Peace*. New Haven, CT: Yale University Press, 1991.

———. *The Rise and Fall of British Naval Mastery*. London: Ashfield Press, 1983.

———. *Strategy and Diplomacy, 1870–1945*. London: Allen and Unwin, 1983.

Keohane, Robert O. *After Hegemony: Cooperation and Discord in the World Political Economy*. Princeton, NJ: Princeton University Press, 1984.

———. "The Demand for International Regimes." *International Organization* 36, no. 2 (1982): 325–55.

———. "International Institutions: Can Interdependence Work?" *Foreign Policy*, no. 110 (1998): 82–194.

———. *Power and Governance in a Partially Globalized World*. London: Routledge, 2002.

Kiesling, Eugenia C. *Arming against Hitler: France and the Limits of Military Planning*. Lawrence: University Press of Kansas, 1996.

Kissinger, Henry A. *American Foreign Policy: Three Essays*. New York: W. W. Norton, 1969.

———. *White House Years*. Boston: Little, Brown, 1979.

Kitching, Carolyn J. *Britain and the Geneva Disarmament Conference*. New York: Palgrave Macmillan, 2003.

Klachko, Mary. "Anglo-America Naval Competition, 1918–1922." PhD Dissertation, Columbia University, 1962.

Knapp, Henry S. "Treaty No. 2 of the Washington Conference." *Political Science Quarterly* 39, no. 2 (1924): 201–17.

Knox, MacGregor, and Williamson Murray, eds. *The Dynamics of Military Revolution 1300–2050*. Cambridge: Cambridge University Press, 2001

Kokoshin, Andrei A. *Soviet Strategic Thought, 1917–1991*. Cambridge, MA: MIT Press, 1998.

Korb, Lawrence J. *The Fall and Rise of the Pentagon: American Defense Policies in the 1970s*. Westport: Greenwood Press, 1979.

Koubi, Vally. "Military Technology Races." *International Organization* 53, no. 3 (1999): 537–65.

Krepinevich, Andrew F. "Cavalry to Computer: The Pattern of Military Revolutions." *National Interest*, no. 37 (1994): 30–43.

Krock, Arthur. *Memoirs: Sixty Years on the Firing Line*. New York: Funk and Wagnalls, 1968.

Kupchan, Charles A., and Clifford A. Kupchan. "Concerts, Collective Security, and the Future of Europe." *International Security* 16, no. 1 (1991): 114–61.

Kurchatov, I. V. "Notes on the Discussion between I. V. Kurchatov, Lead Scientist for the Soviet Nuclear Effort, and Stalin, 01/25/1946." Cold War International History Project, 2007.

Kurth, James R. "A Widening Gyre: The Logic of American Weapons Procurement." In *American Foreign Policy Theoretical Essays*, edited by G. John Ikenberry, 15–38. New York: HarperCollins, 1996.

Kydd, Andrew. "Arms Races and Arms Control: Modeling the Hawk Perspective." *American Journal of Political Science* 44, no. 2 (2000): 228–44.

Labrie, Roger P., ed. *Salt Handbook: Key Documents and Issues 1972–1979*. Washington, DC: American Enterprise Institute, 1980.

Lackoff, Sanford, and W. Erik Bruvold. "Controlling the Qualitative Arms Race: The Primacy of Politics." *Science, Technology and Human Values* 15, no. 4 (1990): 382–411.

LaFeber, Walter. *The Clash: A History of U.S.-Japan Relations*. New York: W. W. Norton, 1997.

Lambert, Nicholas A. *Sir John Fisher's Naval Revolution*. Columbia: University of South Carolina Press, 2002.

Lambeth, Benjamin S. "Deterrence in the MIRV Era." *World Politics* 24, no. 2 (1972): 221–42.

Lapp, Ralph E. *Arms beyond Doubt: The Tyranny of Weapons Technology*. New York: Cowles, 1970.

Lee, William T., and Richard F. Staar. *Soviet Military Policy since World War II*. Stanford, CA: Hoover Institution Press, 1986.

Leebaert, Derek, ed. *Soviet Military Thinking*. London: Allen and Unwin, 1981.

Leffler, Melvyn P. *A Preponderance of Power: National Security, the Truman Administration, and the Cold War*. Stanford, CA: Stanford University Press, 1992.

———. *The Specter of Communism: The United States and the Origins of the Cold War, 1917–1953*. New York: Hill and Wang, 1994.

———. *The Elusive Quest: America's Pursuit of European Stability and French Security, 1919–1933*. Chapel Hill: University of North Carolina Press, 1979.

Leffler, Melvyn P., and David S. Painter, eds. *Origins of the Cold War, an International History*. London: Routledge, 1994.

Levite, Ariel E. "Never Say Never Again: Nuclear Reversal Revisited." *International Security* 27, no. 3 (2002): 59–88.

Levy, Jack. "Declining Power and the Preventive Motivation for War." *World Politics* 40, no. 1 (1987): 82–107.

Levy, Jack S. "Domestic Politics and War." *Journal of Interdisciplinary History* 18, no. 4 (1988): 653–73.

———. "The Offensive/Defensive Balance of Military Technology: A Theoretical and Historical Analysis." *International Studies Quarterly* 28, no. 2 (1984): 219–38.

Liberman, Peter. "The Rise and Fall of the South African Bomb." *International Security* 26, no. 2 (2001): 45–86.

Lieber, Keir A. "Grasping the Technological Peace: The Offense-Defense Balance and International Security." *International Security* 25, no. 1 (2000): 71–104.

———. *War and the Engineers: The Primacy of Politics over Technology*. Ithaca, NY: Cornell University Press, 2005.

Lieberman, Joseph I. *The Scorpion and the Tarantula: The Struggle to Control Atomic Weapons*. Boston: Houghton Mifflin, 1970.

Lilienthal, David E. *The Journals of David E. Lilienthal, Volume II: The Atomic Energy Years, 1945–1950*. New York: Harper and Row, 1964.

Lindsay, James M., and Michael E. O'Hanlon. "Missile Defense after the ABM Treaty." *The Washington Quarterly* 25, no. 3 (2002): 163–76.

Lindsay, James M., Michael E. O'Hanlon, Charles L. Glaser, and Steve Fetter. "Limited National and Allied Missile Defense." *International Security* 26, no. 4 (2002): 190–201.

Lipson, Charles. "International Cooperation in Economic and Security Affairs." *World Politics* 37, no. 1 (1984): 1–23.

Long, Franklin A., and Judith Reppy. *The Genesis of New Weapons: Decision Making for Military R and D*. New York: Pergamon press, 1980.

Lynn-Jones, Sean. "Offense-Defense Theory and Its Critics." *Security Studies* 4, no. 4 (1995): 660–91.

MacDonald, Scot. *The Evolution of Aircraft Carriers*. Washington, DC: Government Printing Office, 1964.

MacGregor, David. "The Use, Misuse, and Non-Use of History: The Royal Navy and the Operational Lessons of the First World War." *Journal of Military History* 56, no. 4 (1992): 603–16.

Mahan, Alfred Thayer. *The Influence of Sea Power upon History, 1660–1783*. Boston: Little, Brown, 1890.

Majeski, Stephen J. "Expectations and Arms Races." *American Journal of Political Science* 29, no. 2 (1985): 217–45.

———. "Technological Innovation and Cooperation in Arms Races." *International Studies Quarterly* 30, no. 2 (1986): 175–91.
Mann, Michael. *The Sources of Social Power: A History of Power from the Beginning to A.D. 1760* Vol. 1. Cambridge: Cambridge University Press, 1986.
Marder, Arthur J. "The Influence of History on Sea Power: The Royal Navy and the Lessons of 1914–1918." *Pacific Historical Review* 41, no. 4 (1972): 413–43.
Marder, Arthur J. *Jutland and after (May 1916–December 1916). From the Dreadnought to Scapa Flow: The Royal Navy in the Fisher Era, 1904–1919.* Vol. 3. London: Oxford University Press, 1966.
Martel, William, and Toshi Yoshihara. "Averting a Sino-U.S. Space Race." *Washington Quarterly* 26, no. 4 (2003).
Martin, Lisa L., and Beth A. Simmons. "Theories and Empirical Studies of International Institutions." *International Organization* 52, no. 4 (1998): 729–57.
Massie, Robert K. *Castles of Steel, Britain, Germany, and the Winning of the Great War at Sea*. New York: Ballantine Books, 2003.
———. *Dreadnought: Britain, Germany, and the Coming of the Great War*. New York: Random House, 1991.
Mastny, Vojtech. *The Cold War and Soviet Insecurity: The Stalin Years*. New York: Oxford University Press, 1996.
———. *Russia's Road to the Cold War*. New York: Columbia University Press, 1979.
Maurer, John. "Arms Control and the Washington Conference." In *The Washington Conference, 1921–22: Naval Rivalry, East Asia Stability and the Road to Pearl Harbor*, edited by Erik Goldstein and John Maurer, 267–93. Portland: Frank Cass, 1994.
MacIsaac, David. "Voices from the Central Blue: The Air Power Theorists," in *Makers of Modern Strategy from Machiavelli to the Nuclear,* edited by Peter Paret. 624–647. Princeton, NJ: Princeton University Press, 1986.
McBride, William M. *Technological Change and the United States Navy, 1865–1945*. Baltimore: Johns Hopkins University Press, 2000.
MccGwire, Michael. *Military Objectives in Soviet Foreign Policy*. Washington, DC: Brookings Institution, 1987.
———. "Soviet Naval Doctrine and Strategy." In *Soviet Military Thinking*, edited by Derek Leebaert, 125–81. London: George Allen and Unvwin, 1981.
McCullough, David. *Truman*. New York: Touchstone, 1992.
McKercher, B. J. C. "The Policies of Naval Arms Limitation in Britain in the 1920s." In *The Washington Conference, 1921–22: Naval Rivalry, East Asia Stability and the Road to Pearl Harbor*, edited by Erik Goldstein and John Maurer, 35–59. Portland: Frank Cass, 1994.
———, ed. *Anglo-American Relations in the 1920s: The Struggle for Supremacy*. Basingstoke: Macmillan, 1991.
———, ed. *Arms Limitation and Disarmament: Restraints on War, 1899–1939*. Westport, CT: Praeger, 1992.
McLellan, David S. *Dean Acheson: The State Department Years*. New York: Dodd, Mead, 1976.
McNamara, Robert S. *The Essence of Security*. New York: Harper and Row, 1969.
McNeill, William H. *The Pursuit of Power: Technology, Armed Force, and Society since A.D. 1000*. Chicago: University of Chicago Press, 1984.
Mearsheimer, John J. "Back to the Future: Instability in Europe after the Cold War." *International Security* 15, no. 1 (1990): 5–56.
———. *Conventional Deterrence*. Ithaca, NY: Cornell University Press, 1983.
———. "The False Promise of International Institutions." *International Security* 19, no. 3 (1994/95): 5–49.
———. "A Realist Reply." *International Security* 20, no. 1 (1995): 82–93.
———. *The Tragedy of Great Power Politics*. New York: W. W. Norton, 2001.
Melhorn, Charles M. *Two Block Fox: The Rise of the Aircraft Carrier 1911–1921*. Annapolis, MD: Naval Institute Press, 1974.
Miller, Edward S. *War Plan Orange: The U.S. Strategy to Defeat Japan, 1897–1945*. Annapolis, MD: Naval Institute Press, 1991.

Millett, Allan R. "Patterns of Military Innovation in the Interwar Period." In *Military Innovation in the Interwar Period,* edited by Williamson Murray and Allan R. Millett, 329–68. Cambridge: Cambridge University Press, 1996.

Millett, Stephen M. "Forward-Based Nuclear Weapons and Salt I." *Political Science Quarterly* 98, no. 1 (1983): 79–97.

Montgomery, Evan Braden. "Breaking out of the Security Dilemma: Realism, Reassurance and the Problem of Uncertainty." *International Security* 31, no. 2 (2006): 151–85.

Murfett, Malcolm H. "Look Back in Anger: The Western Powers and the Washington Conference of 1921–1922." In *Arms Limitations and Disarmament: Restraints on War, 1899–1939,* edited by B. J. C. McKercher, 83–104. Westport, CT: Praeger, 1992.

Murray, Williamson. "Strategic Bombing: The British, American, and German Experiences." In *Military Innovation in the Interwar Period,* edited by Williamson Murray and Allan R. Millett, 96–143. Cambridge: Cambridge University Press, 1996.

Murray, Williamson, and Allan R. Millet, eds. *Military Innovation in the Interwar Period.* Cambridge: Cambridge University Press, 1998.

"National Security Decision Memorandum 49." Nixon Archives (Online), March 27, 1970.

Newhouse, John. *Cold Dawn: The Story of Salt.* New York: Holt, Rinehart and Winston, 1973.

———. "The Missile Defense Debate." *Foreign Affairs* 80, no. 4 (2001).

Nixon, Richard M. *Rn: The Memoirs of Richard Nixon.* New York: Grosset and Dunlap, 1978.

Noel-Baker, Philip. *The First World Disarmament Conference, 1932–1933 and Why It Failed.* Oxford: Pergamon Press, 1979.

Nye, Joseph S. "The Superpowers and the Non-Proliferation Treaty." In *Superpower Arms Control: Setting the Record Straight,* edited by Albert Carnesale and Richard N. Haass, 165–90. Cambridge: Ballinger Publishing, 1987.

O'Connell, Robert L. *Of Arms and Men: A History of War, Weapons, and Aggression.* New York: Oxford University Press, 1989.

———. *Sacred Vessels: The Cult of the Battleship and the Rise of the U.S. Navy.* New York: Oxford University Press, 1991.

O'Hanlon, Michael H., and Mike Mochizuki. "Toward a Grand Bargain with North Korea." *Washington Quarterly* 26, no. 4 (2003).

Olser Hampson, Fen. "Salt I: Interim Agreement and ABM Treaty." In *Superpower Arms Control: Setting the Record Straight,* edited by Albert Carnesale and Richard N. Haass, 65–104. Cambridge: Ballinger Publishing, 1987.

Oye, Kenneth A., ed. *Cooperation under Anarchy.* Princeton, NJ: Princeton University Press, 1986.

Paret, Peter, ed. *The Makers of Modern Strategy from Machiavelli to the Nuclear Age.* Princeton, NJ: Princeton University Press, 1986.

Park, John S. "Inside Multilateralism: The Six-Party Talks." *Washington Quarterly* 28, no. 4 (2005).

Paul, T. V. *Power Versus Prudence: Why Nations Forgo Nuclear Weapons.* Montreal: McGill-Queens University Press, 2000.

Peattie, Mark R. *Sunburst: The Rise of Japanese Naval Air Power, 1909–1941.* Annapolis, MD: Naval Institute Press, 2001.

Pollack, Kenneth, and Ray Takeyh. "Taking on Tehran." *Foreign Affairs* 84, no. 2 (2005).

Posen, Barry. "Command of the Commons: The Military Foundation of U.S. Hegemony." *International Security* 28, no. 1 (2003): 5–46.

———. *The Sources of Military Doctrine: France Britain and Germany between the World Wars.* Ithaca, NY: Cornell University Press, 1984.

Postol, Theodore A. "Troubling Questions About Missile Defense." *Boston Globe,* April 15, 2008, A15.

Potter, William, C. "Coping with MIRV in a Mad World." *Journal of Conflict Resolution* 22, no. 4 (1978): 599–626.

Powell, Robert. "Absolute and Relative Gains in International Relations Theory." *American Political Science Review* 85, no. 4 (1991): 1303–20.

———. "Bargaining Theory and International Conflict." *Annual Review of Political Science,* no. 5 (2002): 1–30.

———. "Review: Anarchy in International Relations Theory: The Neorealist-Neoliberal Debate." *Review of Neorealism and Its Critics* by Keohane, Robert O. in *Neorealism and Neoliberalism: The Contemporary Debate* by Baldwin, David A. *International Organization* 48, no. 2 (1994): 313–44.

———. "War as a Commitment Problem." *International Organization* 60 (2006): 169–203.

Ransom, Harry H. "The Battleship Meets the Airplane." *Military Affairs* 23, no. 1 (1959): 21–27.

Rapoport, Anatol. "Lewis F. Richardson's Mathematical Theory of War." *Conflict Resolution* 1, no. 3 (1957): 249–99.

Rathjens, George W. "The Dynamics of the Arms Race." In *Arms Control: Readings from Scientific American*, 177–87. San Francisco: W. H. Freeman, 1973.

Rattray, Gregory J. *Strategic Warfare in Cyberspace*. Cambridge, MA: MIT Press, 2001.

"A Report on the International Control of Atomic Energy." Washington, DC: U.S. Department of State, 1946.

Resende-Santos, Joao. "Anarchy and the Emulation of Military Systems: Military Organization and Technology in South America, 1870–1930." *Security Studies* 5, no. 3 (1996): 193–260.

———. *Neorealism, States and the Modern Mass Army*. Cambridge: Cambridge University Press, 2007.

Rhinelander, John B. "The Salt I Agreements." In *Salt: The Moscow Agreements and Beyond*, edited by Mason Willrich and John B. Rhinelander, 125–59. New York: Free Press, 1974.

Rhodes, Richard. *The Making of the Atomic Bomb*. New York: Simon and Schuster, 1986.

Rice, Condoleeza. "Salt and the Search for a Security Regime." In *U.S.-Soviet Security Cooperation: Achievements Failures Lessons*, edited by Alexander L. George, Philip J. Farley and Alexander Dallin, 293–306. Stanford, CA: Stanford University Press, 1988.

Richardson, Dick. "The Geneva Disarmament Conference, 1932–1934." In *Decisions and Diplomacy: Essays in Twentieth Century International History*, edited by Dick Richardson and Glyn Stone, 60–82. New York: Routledge, 1995.

Richardson, Lewis F. *Arms and Insecurity*. Pittsburgh, PA: Homewood, 1960.

Richardson, Louise. "The Concert of Europe and Security Management in the Nineteenth Century." In *Imperfect Unions: Security Institutions over Time and Space*, edited by Helga Haftendorn, Robert O. Keohane, and Celeste A. Wallander, 48–80. Oxford: Oxford University Press, 1999.

Richelson, Jeffrey T. *Spying on the Bomb: American Nuclear Intelligence from Nazi German to Iran and North Korea*. New York: W. W. Norton, 2006.

Roberts, Brad, Robert A. Manning, and Ronald N. Montaperto. "China: The Forgotten Nuclear Power." *Foreign Affairs* 79, no. 4 (2000).

Roberts, Chalmers M. "The Road to Moscow." In *Salt: The Moscow Agreements and Beyond*, edited by Mason Willrich and John B. Rhinelander, 3–33. New York: Free Press, 1974.

Roco, Mihail C., and William S. Bainbridge, eds. *Societal Implications of Nanoscience and Nanotechnology*. Norwell, MA: Kluwer Academic Publishers, 2001.

Roe Smith, Merritt, and Leo Marx, eds. *Does Technology Drive History? The Dilemma of Technological Determinism*. Cambridge, MA: MIT Press, 1994.

Rosecrance, Richard. *Rise of the Trading State: Commerce and Conquest in the Modern World*. New York: Basic Book, 1987.

Rosecrance, Richard, and Arthur Stein, eds. *The Domestic Bases of Grand Strategy*. Ithaca, NY: Cornell University Press, 1991.

Rosen, Stephen Peter. *Winning the Next War: Innovation and the Modern Military*. Ithaca, NY: Cornell University Press, 1991.

Rosenberg, David Alan. "The Origins of Overkill: Nuclear Weapons and American Strategy, 1945–1960." *International Security* 7, no. 4 (1983): 3–71.

Roskill, Stephen. *Naval Policy between the Wars, Volume 1: The Period of Anglo-American Antagonism, 1919–1929*. New York: Walker and Company, 1968.

Ross, Steven T. *American War Plans 1890–1939*. London: Frank Cass, 2002.

Rothenberg, Gunther E. "Moltke, Schlieffen, and the Doctrine of Strategic Envelopment." In *Makers of Modern Strategy from Machiavelli to the Nuclear Age*, edited by Peter Paret, 296–325. Princeton, NJ: Princeton University Press, 1986.

Sagan, Scott D. "1914 Revisited: Allies, Adversaries, and Instability." *International Security* 11, no. 2 (1986): 151–75.

———. "How to Keep the Bomb from Iran." *Foreign Affairs* 85, no. 5 (2006).

———. "Why Do States Build Nuclear Weapons?: Three Models in Search of a Bomb." *International Security* 21, no. 3 (1996): 54–86.

Sample, Susan G. "Arms Races and Dispute Escalation: Resolving the Debate." *Journal of Peace Research* 34, no. 1 (1997): 7–22.

———. "Furthering the Investigation into the Effects of Arms Buildups." *Journal of Peace Research* 35, no. 1 (1998): 122–26.

Schelling, Thomas C., and Morton H. Halperin. *Strategy and Arms Control*. Washington, DC: Pergamon-Brassey, 1985.

Schencking, J. Charles. *Making Waves: Politics, Propaganda, and the Emergence of the Imperial Japanese Navy, 1868–1922*. Stanford, CA: Stanford University Press, 2005.

Schwartz, David N. "Past and Present: The Historical Legacy." In *Ballistic Missile Defense*, edited by Ashton B. Carter and David N. Schwartz, 330–49. Washington, DC: Brookings Institution, 1984.

"Secretary Rogers Urges Support of the ABM Treaty and Interim Agreement on Strategic Offensive Weapons." *Department of State Bulletin* 67, no. 1724 (1972).

Sherwin, Martin J. "The Atomic Bomb and the Origins of the Cold War." In *Origins of the Cold War: An International History*, edited by Melvyn P. Leffler and David S. Painter, 77–94. London: Routledge, 1994.

———. "The Atomic Bomb and the Origins of the Cold War: U.S. Atomic-Energy Policy and Diplomacy, 1941–45." *American Historical Review* 78, no. 4 (1973): 945–68.

———. *A World Destroyed: Hiroshima and the Origins of the Arms Race*. New York: Vintage Books, 1987.

Shils, Edward. "The Failure of the United Nations Atomic Energy Commission: An Interpretation." *University of Chicago Law Review* 15, no. 4 (1948): 855–76.

Shimshoni, Jonathan. "Technology, Military Advantage, and World War I: A Case for Military Entrepreneurship." *International Security* 15, no. 3 (1990): 187–215.

Sienkiewicz, Stanley. "Soviet Nuclear Doctrine and the Prospects for Strategic Arms Control." In *Soviet Military Thinking*, edited by Derek Leebaert, 73–91. London: George Allen and Unwin, 1981.

Singh, Jaswant. "Against Nuclear Apartheid." *Foreign Affairs* 77, no. 5 (1998).

Smith, Gerard C. *Doubletalk: The Story of Salt I*. Lanham, MD: University Press of America, 1985.

Snyder, Glenn H. *Deterrence and Defense: Toward a Theory of National Security*. Princeton, NJ: Princeton University Press, 1961.

———. "Process Variables in Neorealist Theory." *Security Studies* 5, no. 3 (1996): 167–92.

Snyder, Jack. *Myths of Empire: Domestic Politics and International Ambition*. Ithaca, NY: Cornell University Press, 1991

———. "Civil-Military Relations and the Cult of the Offensive, 1914 and 1984" *International Security* 9, no. 1 (1984).

The Soviet Military Technological Challenge. Washington, DC: Center for Strategic Studies, Georgetown University, 1967.

Spanier, John W., and Joseph L. Nogee. *The Politics of Disarmament: A Study in Soviet-American Gamesmanship*. New York: Praeger, 1962.

Sprout, Harold. "Changing Power Relations in the Pacific." *Annals of the American Academy of Political and Social Science* 215 (1941): 107–14.

Sprout, Harold, and Margaret Sprout. *The Rise of American Naval Power, 1776–1918*. Princeton, NJ: Princeton University Press, 1946.

———. *Toward a New Order of Sea Power: American Naval Policy and the World Scene, 1918–1922*. New York: Greenwood Press, 1976.

Spruyt, Hendrick. *The Sovereign State and Its Competitors* Princeton, NJ: Princeton University Press, 1996.

Stein, Arthur. *Why Nations Cooperate, Circumstance and Choice in International Relations*. Ithaca, NY: Cornell University Press, 1990.

Stevens, Sayre. "The Soviet Bmd Program." In *Ballistic Missile Defense*, edited by Ashton B. Carter and David N. Schwartz, 182–220. Washington, DC: Brookings Institution, 1984.
Stimson, Henry L., and McGeorge Bundy. *On Active Service in Peace and War*. New York: Harper, 1948.
"Strategic Arms Limitation Agreements Transmitted to Congress." *Department of State Bulletin* 67, no. 1723 (1972).
Subcommittee of House Committee on Appropations. *Naval Appropriation Bill, 1922*, 66th Cong., 3rd Sess, 1921.
Sullivan, Brian R. "Italian Naval Power and the Washington Conference of 1921–22." In *The Washington Conference, 1921–22: Naval Rivalry, East Asian Stability and the Road to Pearl Harbor*, edited by Erik Goldstein and John Maurer, 220–48. Essex: Frank Cass, 1994.
Suri, Jeremy. "Explaining the End of the Cold War: A New Historical Consensus?" *Journal of Cold War Studies* 4, no. 4 (2002): 60–92.
Talbott, Strobe. "Dealing with the Bomb in South Asia." *Foreign Affairs* 78, no. 2 (1999).
Talensky, Nikolai A. "Anti-Missile Systems and Disarmament." In *The Military-Technical Revolution: It Impact on Strategy and Foreign Policy*, edited by John Erickson, 219–28. New York: Praeger, 1966.
Taliaferro, Jeffrey W. "Security Seeking under Anarchy: Defensive Realism Revisited." *International Security* 25, no. 3 (2000/01): 128–61.
———. "State Building for Future Wars: Neoclassical Realism and the Resource-Extractive State." *Security Studies* 15, no. 3 (2006): 464–95.
Tammen, Ronald L. *MIRV and the Arms Race: An Interpretation of Defense Strategy*. New York: Praeger, 1973.
Taubman, William. *Stalin's American Policy: From Entente to Detente to Cold War*. New York: W. W. Norton, 1982.
Thatcher, Ian D. "Brezhnev as Leader." In *Brezhnev Reconsidered*, edited by Edwin Bacon and Mark Sandle, 22–37. Hampshire: Palgrave Macmillan, 2002.
Thayer, Bradley A. "The Political Effects of Information Warfare: Why New Military Capabilities Cause Old Political Dangers." *Security Studies* 10, no. 1 (2000): 43–85.
Thee, Marek. *Military Technology Military Strategy and the Arms Race*. New York: St. Martin's Press, 1986.
Till, Geoffrey. "Adopting the Aircraft Carrier: The British, American, and Japanese Case Studies." In *Military Innovation in the Interwar Period*, edited by Williamson Murray and Allan R. Millett, 191–226. Cambridge: Cambridge University Press, 1996.
Tilly, Charles. *Coercion, Capital, and European States, Ad 990–1992*. Cambridge: Blackwell, 2002.
Tolles, William M. "National Security Aspects of Nanotechnology." In *Societal Implications of Nanoscience and Nanotechnology*, edited by Mihail C. Roco and William S. Bainbridge. Norwell, MA: Kluwer Academic Publishers, 2001.
Trachtenberg, Marc. *A Constructed Peace: The Making of the European Settlement, 1945–1963*. Princeton, NJ: Princeton University Press, 1999.
———. *History and Strategy*. Princeton, NJ: Princeton University Press, 1991.
———. "Strategic Thought in America, 1952–1966." *Political Science Quarterly* 104, no. 2 (1989): 301–34.
Truman, Harry S. *Memoirs of Harry S. Truman: 1945 Year of Decisions*. Vol. 1. New York: Doubleday, 1955.
———. *Memoirs of Harry S. Truman: 1946–52 Years of Trial and Hope*. New York: Konecky and Konecky, 1955.
Van Creveld, Martin. *Technology and War: From 2000 B.C. To the Present*. New York: Free Press, 1989.
Van Evera, Stephen. *Causes of War: Power and the Roots of Conflict*. Ithaca, NY: Cornell University Press, 1999.
———. "The Cult of the Offensive and the Origins of the First World War." *International Security* 9, no. 1 (1984): 58–107.
———. *Guide to Methods for Students of Political Science*. Ithaca, NY: Cornell University Press, 1997.

———. “Offense, Defense and the Causes of War.” *International Security* 22, no. 4 (1998): 5–43.

Vigor, P. H. *The Soviet View of Disarmament*. New York: St. Martin’s Press, 1986.

Vinson, J. Chal. “The Drafting of the Four-Power Treaty of the Washington Conference.” *Journal of Modern History* 25, no. 1 (1953): 40–47.

Wallace, Michael D. “Arms Races and Escalation: Some New Evidence.” *Journal of Conflict Resolution* 23, no. 1 (1979): 3–16.

———. “Comments on the Articles by Sample and Diehl and Crescenzi.” *Journal of Peace Research* 35, no. 1 (1998): 119–21.

Waller, Douglas C. *A Question of Loyalty: Gen. Billy Mitchell and the Courtmartial That Gripped the Nations*. New York: HarperCollins, 2004.

Walt, Stephen M. *The Origins of Alliances*. Ithaca, NY: Cornell University Press, 1984.

———. *Taming American Power: The Global Response to U.S. Primacy*. New York: W. W. Norton, 2005.

Waltz, Edward. *Information Warfare Principles and Operations*. Boston: Artech House, 1998.

Waltz, Kenneth M. *Theory of International Politics*. New York: McGraw-Hill, 1979.

Waltz, Kenneth N. “The Emerging Structure of International Politics.” *International Security* 18, no. 2 (1993): 44–79.

Webber, Steve. *Cooperation and Discord in U.S.-Soviet Arms Control*. Princeton, NJ: Princeton University Press, 1991.

Weiner, Stephen. “Systems and Technology.” In *Ballistic Missile Defense*, edited by Ashton B. Carter and David N. Schwartz, 49–99. Washington, DC: Brookings Institution, 1984.

Wertheimer, Mildred S. “The League of Nations and the Prevention of War,” *Foreign Policy Reports* 6:11. New York: Foreign Policy Association, 1930, pp. 206–24.

Wheeler-Bennett, John W. *Disarmament and Security since Locarno, 1925–1931*. New York: Howard Fertig, 1973.

———. *The Pipe Dream of Peace: The Story of the Collapse of Disarmament*. New York: William Morrow, 1935.

Wheeler, Gerald E. *Prelude to Pearl Harbor: The United States Navy and the Far East, 1921–1931*. Columbia: University of Missouri Press, 1968.

Williams, Bejamin H. *The United States and Disarmament*. New York: Whittlesey House, 1931.

Willrich, Mason. “Salt I: An Appraisal.” In *Salt: The Moscow Agreements and Beyond*, edited by Mason Willrich and John B. Rhinelander, 256–76. New York: Free Press, 1974.

Willrich, Mason, and John B. Rhinelander, eds. *Salt: The Moscow Agreements and Beyond*. New York: Free Press, 1974.

Wolfe, Thomas W. *Soviet Strategy at the Crossroads*. Cambridge, MA: Harvard University Press, 1964.

Woodward, David. “Limitation of Land Armaments,” *Foreign Policy Reports* 6:2. New York: Foreign Policy Association, 1930, pp. 17–35.

———. “Limitation of Air Armaments,” *Foreign Policy Reports* 6:17. New York: Foreign Policy Association, 1930, pp. 295–313.

Wright, Quincy. “The Washington Conference.” *American Political Science Review* 16, no. 2 (1922): 285–97.

Yanarella, Ernest J. *The Missile Defense Controversy: Strategy, Technology, and Politics, 1955–1972*. Lexington: University Press of Kentucky, 1977.

York, Herbert F. “ABM, MIRV, and the Arms Race.” *Science* 169, no. 3942 (1970): 257–60.

———. *Race to Oblivion: A Participant’s View of the Arms Race*. New York: Clarion, 1970.

Zaborski, Jason. “Deterring a Nuclear Iran.” *Washington Quarterly* 28, no. 3 (2005).

Zakaria, Fareed. *From Wealth to Power: The Unusual Origins of America’s World Role*. Princeton, NJ: Princeton University Press, 1998.

———. “Review: Realism and Domestic Politics: A Review Essay.” *Review of Myths of Empire: Domestic Politics and International Ambition* by Snyder, Jack. *International Security* 17, no. 1 (1992): 177–98.

Zhukov, Georgii Konstantinovich. *The Memoirs of Marshal Zhukov*. New York: Delacorte Press, 1971.

Zubok, Vladislav, and Constantine Pleshakov. *Inside the Kremlin's Cold War: From Stalin to Khrushchev*. Cambridge, MA: Harvard University Press, 1996.

Index

About the Author

David W. Kearn Jr. is assistant professor in the Department of Government and Politics at St. John's University. His research interests include international relations theory, U.S. foreign policy, military innovation, and arms control. Dr. Kearn received his B.A. from Amherst College, a Master of Public Policy (MPP) from the John F. Kennedy School of Government at Harvard University, and a PhD in Foreign Affairs from the University of Virginia.

www.ingramcontent.com/pod-product-compliance
Ingram Content Group UK Ltd.
Pitfield, Milton Keynes, MK11 3LW, UK
UKHW012037150926
475560UK00001B/62

9 780739 189436